# Time Dancing
## A Galactic Journal

Written by

Paula A Adkins

# Ackowledgement

I would like to acknowledge my husband Stan for his ability to always make me feel I could do anything I put my mind to. He was the love of my life.

I would also like to thank my parents for giving me a curious mind and a quest for knowledge when it came to thinking outside the box. They inspired my fascination with all extraordinary, outrageous and other worldly things. I would also like to thank my little brother who helped me remember our abduction event.

With my family at my side, I wrote the story of my life in a science fiction framework. These wonderful family members have all passed away and are now my ghost writes and muses.

# Dedication

This novel is dedicated to all of the star seeds that occupy this blue pearl called Earth.

We are all on a ship floating among the stars. Our tiny world is filled with diversity, both on the ground and under the sea. I dedicate this novel to Earthlings in hopes we can find a path for peace.

Paula A Adkins #9585
Time Dancing a galactic journal

# Prologue

Time, like the vehicle driving our lives, is thought to be solidly locked onto a one-way road without off-ramps or turn-a-rounds. It is always moving forward with only memories of the past and hope for the future. It is like being on a road trip without a map or a guaranteed destination.

The archeology of past civilizations has yet to fully reveal the mysteries of our history. Those anomalies of time's mysteries continue to change the scope of Earth's real history. Adventurers are discovering many different cultures, religious relics, and potential alien involvement, which might be proving times paradoxes that are being found among the ruins of ancient civilizations. It is those clues left by Earth's mysterious advanced civilizations which contains hidden messages and warning of time and space and parallel universes.

Open up your mind and become the archeologist of the past, the present and the future. These facts are brought to you by blending solid history with the notion that what is written in science fiction format is no more or less factual than the flat Earth theory. Think outside the box and journey with me through imagination and reality from planet Earth and beyond.

Welcome to the concept of time dancing. Join me on my journey though the many folds of times.

All historical researched history and people are based on fact and will be marked with a *. Characters created to advance the story are fictional. Any accidental connection to real people's names or occupations, living or not within the plot of the story, are coincidence and nothing more.

# CHAPTER 1

## Mystery Quest

The news feed on the Federation Building in downtown Phoenix flashed on as Professor Ryan made his way to the Smithsonian Board of Regent's corporate office.

> October 10, 2050: **Smithsonian Archeologist to excavate part of the Grand Canyon's South Rim**. Environmental groups up in arms. Professor Ryan of the Smithsonian's Archeologist Research and Exploration Unit will be heading the archeological dig…more news to follow….

Jeff Ryan had been investigating hidden information on the mythological Seven Cities of Cibola. It had been his passion for as long as he could remember. His Master's Thesis had been called "*The lost cities of Cibola, the Egyptian Connection.*"

Jeff had discovered an old newspaper archive dated 1909. This article reported the discovery of a rock-cut vault inside the sheer walls of the Grand Canyon early in his research of myths surrounding this impassable cave. Jeff, an archeology Professor at Arizona State University, was an avid historian. He was fascinated with ancient Hopi myths and their connection with sky gods. It was not unlike the Sumerian texts of antiquity. The odd article added the Egyptian connection, much to his fascination.

The erosion of the South Rim of the Grand Canyon had revealed a cut stone effigy that remarkably resembled the image of the sun god, Ra. The professor had spent almost a year working on paperwork and red tape attempting to get the World Park Systems and the Smithsonian to first allow the excavation and, second, to release what he could only assume to be known information hidden by the government for some unknown reason.

Jeff had seriously wondered why the Smithsonian's Archeology Department had originally hidden the relics retrieved in the first exploration of those caverns. This, of course, required verification that the article had merit. It was a long shot at best, but it coincided with his research. If the article had merit, what information had they felt so dangerous that the knowledge would cause concern to the general public?

Professor Jeff Ryan brushed his curly hair away from his deeply tanned, chiseled face. He glided his rough hands down his dress shirt and pants. Even though he was in what most would call business casual, his bulging muscular build and tall stance told most of the business people inside the front information center that this gentleman was not a business man, scientist, or technologist but an adventurer.

Jeff approached the main desk.

"Hello, I am Professor Ryan here to see the Board of Regents of the Smithsonian. I am expected."

The young woman behind the computer screen looked up with a smile and simply told him to go to the twelfth floor, Room

201. The young woman kept her gaze on this ruggedly handsome man as he walked to the lift and up to the Regents Office.

Jeff was concerned that the dig was to be canceled due to the outrage of the Environmental Protection Agency's concerns for the continued beauty of the Grand Canyon. As he opened the office door, a small framed man in a long white tunic looked up from his work and gestured him to sit down. Jeff wondered where the rest of the board was, but he dutifully sat.

"Professor Ryan, I am pleased to give you the official permission for your upcoming dig, but I must tell you that there are some conditions that you must agree to. One, any information that is found that is proof of a time paradox must be instantly sealed in whatever area it is found. Second, no information gleaned which may indicate to the public of altered time or altered reality must be forever sealed and NOT reported or repeated. Is this understood?"

"Time what? We already use time travel in public schools… what altered reality are you speaking of?"

"You will know it if you find it… you are going where G.E. Kinkaid traveled in 1906. One of the Smithsonian Archeologists, S.A. Jordon, retraced Kinkaid's expedition in 1909 and provided a detailed map of over twelve arms to a cavern system that drops down over one thousand feet. Kinkaid vanished after he found this remarkable cliff dwelling. The information Professor Jordon found drove him crazy, and he was finally sent to a mental institution, where he spent the rest of his life. Some information is so controversial that it is better left a secret. This dig is to further advance your very interesting Master Thesis, "*The lost cities of Cibola, the Egyptian Connection,*" and nothing more. We, at the Smithsonian, believe that inside the inner cities is a vast archival library that has been hidden for thousands of years. Some of the relics are yet to be found, and we believe that there are many golden statues and stashes of gold that the Federation is in need of for expanding our holdings in the galaxy. Your permission comes with a cost…the finds remain the exclusive property of the Smithsonian… is that agreeable to you?"

"I suppose it will have to be…won't it." Jeff placed his hand on the document in front of him, and a pinprick pierced his lower palm. His blood drop and palm print were recorded for verification… the deal was signed.

Jeff left with a sinking feeling as he mulled over the time paradox comment…what had the Smithsonian been hiding for over one hundred and forty years. What historical information could be so harmful as to cause an open and free world to cloister away historic records? He would have to make it his mission to discover what the mystery was.

# CHAPTER 2

## Mythological Cities of Gold

As Professor Ryan arrived at the University parking area, he noticed his students amassed and ready to travel. They must have been excited as they were an hour early. Professor Ryan had more volunteer students than he needed from his advanced archeology class and had to whittle the volunteers down to eleven. He had four avid rock climbers in his advanced class, with vast experience in traversing thousand-foot sheer rock faces… he would need them.

The rest of the students are very bright, fit and eager up-and-coming archeologists in their own right. They boarded the fully equipped van and headed to the South Rim of the Grand Canyon.

His rock climbers had been given the option to come or stay and were told of the challenges they alone would have to face. The effigy, as well as the chisel-cut stone entrance to the cave and to the first wagon wheel-shaped cavern system, was 1,486 feet from the top mesa and over 2,000 feet from the canyon below.

It would have to be decided which way would be the best to traverse the wall, down a sheer cut rock face or up a partially rocky canyon with a small sandy beach by Crystal Creek. This area between Desert View and Moran Point, named the Haunted Canyon by the Hopi, had been off-limits for over 140 years… until today.

While the van drove up towards the entrance of the Grand Canyon, the students took a vote to decide who would get access to the equipment in the van. Since it would be easier to drop down to the cavern entrance. Once the plan of attack was finalized, Jeff explained to his students that any finds that seemed out of place should be reported directly to him. Those items were not to be video recorded but viewed by his eyes only.

Once he had cleared them, those items that could be removed or the video recording could be analyzed that evening at camp using the equipment on board the van. There was a lot of complaining over the "professor's eyes only ruling," but Jeff explained that the restriction was placed on him by the Board of Regents with the Smithsonian, who was funding the excavation.

The first few days, the students set up the camp. The conscious of the group was to approach the cave from above. The students built a large pulley system to be used to transport the students and equipment into the cave entrance. The rock climbers had the hardest job originally, as it was the responsibility of those four athletic students to anchor the pulley system to the cave entrance. They actually shouted exuberant yells when they swung inside the cave entrance and viewed the golden-colored interior walls.

The night before the group was to begin the exploration of the interior cavern system, Jeff explained to the students what the main mission was— both his and the Regent's.

"Sid already told you that he thought he saw gold inside the cavern… it is true. The reports from the archives say that the golden walls are part of the seven cities of Cibola, part Spanish Conquistador's myth and part of the Hopi Myth. All this corresponded with the hidden report from Professor Jordon, who traversed the wall and entered the cavern in 1909. We are to gather golden artifacts, statues and jewelry for the Smithsonian. I would like to find the ancient library that I believe is hidden within the inner cities.

"I believe, like the ancient Alexandria Library of early Egyptian myth, there is an equally valuable stash of early Earth history to be found…you know, those lost years of technological wonders of the pyramid builders that pre-dated our written history, so keep your eyes open for that. We will *only* record at this point, so do not move the artifacts until we have the Regents examine our finds. Get a good night's sleep. Tomorrow will be a long day."

Jeff sat silently and looked up at the night sky. Thousands of stars shone, and the edge of the Milky Way was so bright that it made the cosmos feel as though it were within his reach. He raised his arms towards the sky, and a shooting star streaked past his open arms. *Good sign*, he thought to himself and retired for the night.

The next morning the eleven students readied themselves for the drop into the cave entrance. Sid and the rock climbers pivoted over the edge of the canyon walls hanging like spiders over half a mile-high drop, two on one side of the pulley and two on the other.

Their job was to keep the pulley from swinging in the wind as it moved down the sheer canyon walls. They were securely anchored, one at 400 feet and the other at 700 feet. Safety harnesses kept them from also swaying with the prevailing wind. The magnetic pitons were non-evasive, which would have possibly made the environmentalist a little less concerned with the damage to the wall.

Jeff was the first person to be attached to the pulley system, as he would be the controlling counter balance at the other end. He also wanted to make sure it was safe for his students to enter this cavern system that had not seen a human in one hundred and forty years. There was no telling what he would find occupying the inner spaces now. The drop was breathtaking as Jeff resisted looking at the ground.

He was an archeologist first and a speleologist second, not a mountain climber. Caves, walls, dirt and earth. All these elements were better than the open air and long drops. It wasn't that he had a fear of heights but a fear of falling from those heights. Finally, he kicked off the sheer rock face swinging out and into the cave entrance.

He anchored the double ropes to the interior floor with metal pitons and radio up to the mesa to come on down. The remaining seven students used the pulley system, and the rock climbers slowly lowered themselves to the remaining feet and swung into the cave. Professor Ryan had already placed the powerful laser luminescent lighting ropes around the interior of the main cavern. The walls were indeed shining with a golden cast, and ancient

Hieroglyphs and petroglyphs adorned the inner walls. Jeff marveled at the combination of the two unrelated wall art.

Jeff pulled out a rolled-up paper and laid it on the ground.

"Let's make a plan to adventure deeper into the caves. I think we should stay in a pairing system and report to each other every fifteen minutes. Keep your GPS mapping systems on, as well as your communication devices. These caverns are shaped like a wagon wheel, but they continue to a greater distance than what has been explored before. We now have the technology to be able to light our way in the darkest of areas, and I want you to video record all that you are viewing. I want you each to record the image of this antique map. It may not be accurate, but it is all we have."

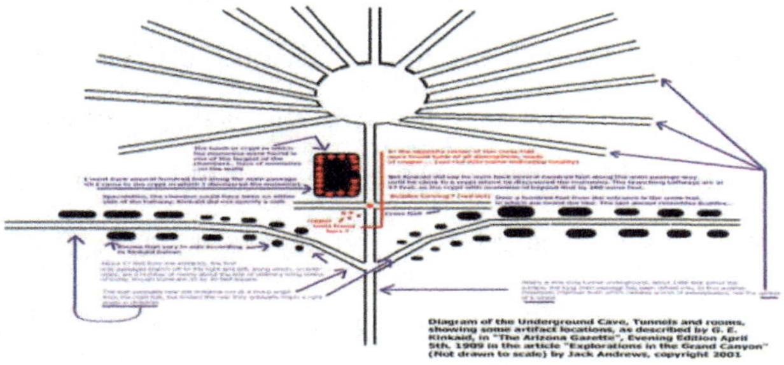


Phoenix Gazette 04/5/1909

We are currently moving towards the center circle, with twelve directions to choose from. I think we should start with the west corridors first. I will pick Sid as my partner, and the rest of you will pair up. Sid and I will take the last arm to the west. Remember, take care, be safe, record and do not remove anything. Report every fifteen minutes. My watch tells me it is 7:45. I want to hear your voices starting at 8 a.m."

The group was off to explore. Jeff and Sid took the last corridor. Sid was recording almost every hieroglyph he came across while Jeff took samples of the wall material at the base as well as some of the already faded paint from the artistic Egyptian designs that decorated the twelve-foot-wide corridor. After twelve minutes, Jeff and Sid arrived at what looked like an opening to the corridor. There was a foul smell, almost like a snake enclosure; Jeff put his arm up to stop Sid from entering while he opened up the light beam and searched the floor for what he could only imagine as a rattlesnake nest of a magnitude to emit that much odor.

The shed skin of snakes was everywhere, but no signs of the serpents in question. They slowly moved into the opening. As Jeff panned the light wand upwards towards the wall, a golden face appeared, and the communication device beeped. They both jumped out of shock. Sid laughed, and Jeff let out a gush of air.

"News…" Jeff said as he heard laughter and yelling on the other end. The first pair of students to report in was so loud their communication was garbled.

"Slow down, take a breath…what do you have to report?"

"We have more than just hieroglyphics! We have a giant mural of what I think is the god Ra with two women on either side of him. It looks like they are offering him food. We have it recorded.

It's beautiful and not faded in any way. Looks like it was painted yesterday."

"Which corridor are you monitoring?"

"The fifth one, the one above you."

Jeff again panned the area where he though he saw a face shining, and indeed, what looked very much like a sitting Buddha was straight ahead. Sid moved closer to the statue, which was over ten feet tall and began videotaping from every angle.

The communication device beeped again. Each pair of students reported remarkable images of ancient gods from every corner of the world, from the Aztec Empire to the Mayan, Egyptian, India and African deities. This was the archeologist find of the century! The key to the library of knowledge. The archives of ancient spirituality are held within the spokes of the west side corridors. Every fifteen minutes, the students continued to check-in. It was now approaching 4 p.m., and Jeff decided it was time to return to the main cavern and to the surface to review their finds. Tomorrow they would document the east side corridors.

By the time they finally got to camp, it was dusk, and the first star twinkled above. The van was teaming with students, each talking over the next in excited communications of what they had recorded within the corridors. Jeff didn't enter the van. He sat outside contemplating their next move. Other than the walls, which were far too valuable historically, the only gold appeared to be the

Buddha statue, which would be too heavy for this expedition to remove for the Smithsonian.

Jeff figured the Regents could worry about the logistics of the move. The golden walls. Well, that was another matter. It would have to be a heated debate to save the hieroglyphics and the mural for the museum, gold or not. Jeff had the last of his coffee and headed for the tent, the students were too excited to go to bed, and he didn't blame them. He was twice their age, and he needed the rest.

The early morning sun woke Jeff from a deep sleep, and suddenly he heard what sounded like a shuttle craft hovering over the camp. He pulled his pants on a dashed out of the tent. The craft landed fifty feet away from the edge of the canyon, but the dirt still rose from the mesa. As the dust cleared, Jeff could make out several men dressed in the librarian tunic uniforms of the Smithsonian. He walked towards them, straightening his curly mass of brown hair while yawning. He couldn't help but wonder about what they wanted.

"Professor Ryan, have you made any progress with the dig?"

"This isn't instant pudding. We have just begun. The first day was yesterday, and I already had the students forward the video to your office last night. Then, what's the rush for?"

"The Environmental Protection Agency has gone to the Federation to stop your dig. We're just hoping you have found something that will prove to them this venture is worthwhile. Have you?"

"We found what appears to be part of a greater library of Earth's spiritual traditions, but we haven't really had time to fully inspect each corridor or go deeper into the inner city, where I hope to find Cibola. Is there any way you can intervene on our behalf to give us at least thirty days?"

"We will try. Do you have any artifacts we can take to the Federation?"

"No, not unless you are willing to open up the side of the canyon wall and remove the ten-foot-tall golden Buddha. That would really piss those environmentalists off for sure."

"Buddha...what is that statue doing in the desert southwest?"

"Library, museum of the past; everything is unknown at this point. Please give us the opportunity to venture deeper into the inner caverns. Please!"

By this time, the students had all left their tents and were standing in a semi-circle behind the professor. They had expressions of anger, concern, and some occasionally yawned. Jeff looked at his students and knew they had stayed up half the night discussing their adventure. He had hoped they would continue to have this once-in-a-lifetime opportunity to be on the cutting edge of discovery. Bureaucracy, he thought, never changes, no matter how peaceful the galaxy is.

Jeff turned to return to his tent when Sally, one of his rock climbers, stood staring at his muscular nude chest. He was at first flattered, then embarrassed and dashed into his tent to find his shirt and hat for the morning drop into the cave.

After everyone was fed and the coffee drained, they began their descent into the cave. This morning, they would be exploring the east side corridors, skipping the first ones. The west side number one corridor had petroglyphs of South American mythology, and it was assumed the same would hold true for the first east side corridor. Jeff and Sid would traverse the center-long corridor, which had been left unexplored by Professor Jordon. Jordon had reported that there appeared to be a large crevice that was not lit with the bright lighting they carried with them.

He also reported a deadly 'reptilian smell' emanating from the pitch-black cavern. Jeff had the idea that this may have been the main entrance to the inner civilizations reported to be one of the seven lost cities of Cibola. Again, the rules were to report every fifteen minutes and to record what they saw.

Sid turned out to be an excellent videographer which gave Jeff the opportunity to examine areas of the corridor and insets which could contain papyrus or other jars of knowledge from the past. He, unlike his students, didn't have the compulsion to leave things where they lay; he would collect if he thought the Regents would censor the material.

After two hours had gone by, Sid and Jeff finally came to a main room which was seven-hundred feet by one-hundred and twenty-five feet. It contained what looked like cooking utensils, pottery, and water jugs, which were now filled with dust. The walls were artfully decorated with murals depicting the desert as well as what appeared to be the Giza plateau in the distance. With the ultra-luminescent lighting, it looked to be a mural that would provide the optical illusion that the occupants were eating outside.

Sid spent at least an hour video recording every inch of this remarkable room while Jeff slipped away into a side room that appeared to have been an apartment or family dealing of some kind. At the back of the room, in a corner, Jeff saw what appeared to be a wooden box with decorations on it. It wasn't large enough to be a mummy but an interesting find. Sid had noticed the Professor's absence and called out. Jeff yelled back, "Just in the next room; keep doing what you're doing. I'll be there in just a moment."

As Jeff brushed off the dust with his round camel hair brush, the chest became clear. Jeff stood up in shock. "What the hell."

"You okay, Professor?" Sid called out.

"Just tripped over my own feet. I'll be there in a minute."

The chest had a pink flop-eared bunny painted on the outside. It looked a lot like his mother's toy box she had attempted to give him when he was a child. His mother's toy box was the same shape, the same latch but had an odd purple lizard on the outside. The toy chest was strange…out of place…out of time. He gently opened the metal clasp on the toy box and dusted the dirt away from what looked to be heavy parchment and a large leather-bound book.

After cleaning what looked like antique parchment paper, the image appeared - a faded paper-bound book. This small paper book was a 1953's "Snow White and the Seven Dwarves." The paper was old and fragile, the cardboard cover faded and dog-eared. Jeff slid it into his pack gently. The larger leather-bound book had a locked clasp on it, and he was fearful of disturbing it here. He grabbed a handful of dirt and quickly covered the toy box as if it hadn't been discovered. He slid the leather-bound book into his pack as well. He would inspect this in his tent tonight. What was a wooden box, which obviously belonged to a child of the 20th century, doing in a prehistoric cavern? The more he contemplated the find, the more he wondered if this was the time thing the Regent had talked about.

Jeff's curiosity was burning. Time couldn't move fast enough now until he could inspect this document. Jeff pondered that thought and then dismissed it. What earth-shattering information could be deemed so secret in a child's book and a toy chest?

Jeff re-entered what Sid had decided to call the main dining room. They had inspected at least fifty rooms branching off this one area but hadn't ventured any deeper into the cavern after returning from the peculiar black cavern at a dead end in the long corridor.

Jeff speculated that this city could have accommodated at least fifty-thousand people if the rest of the domiciles were arranged similarly. This information was featured in a 1909 article, and his estimate corroborated that information. Professor Ryan was as excited as his students, who reported gold statues scattered in each of the corridors along with evidence of mummification. The students remarked on the striking resemblance to museum artifacts from Egypt.

Finally, at 4:45, the students gathered in the main room for extraction from the inner rooms and corridors. The day had been quite eventful, and Jeff's curiosity was piqued.

Back at camp, the students uploaded their video to the Smithsonian and spent the rest of the evening reviewing what the other group had observed. Jeff retreated to his tent with a sandwich and a cup of coffee. He first inspected the child's book for any clues about its owner. Something caught his attention - a name, possibly, in what looked like Arabic, penned on the left-hand inside cover - سكايلر. He snapped a copy of the word into his data card and requested a translation.

The card interpreted it as "ancient Arabic...proper name...Skylar". This was peculiar, more than just peculiar. He had reports of ancient artifacts from around the globe, but here was a contemporary English book with a foreign inscription. The meaning eluded him.

He was eager to open the leather-bound book but decided to take a moment to eat and reflect. After his meal, he donned a white linen glove, carefully unlocked the book and opened it. His eyes widened. The contents were not what he had expected to see.

How this document had ended up in what seemed like a family unit within an ancient Anasazi Cliff dwelling was baffling. The find represented an unthinkable paradox - a clash of timelines spanning centuries. Jeff considered a possibility, then dismissed it - maybe a family of past mountain climbers had found this cavern dwelling and moved in. But how would such a family have managed the climb or drop into the cavern, let alone live without light, food, or water? And why would they bring along something as impractical as a wooden toy box?

Only the toy box had been found in that dwelling. There were no bones, no other signs of occupancy apart from petroglyphs. An animal on one side and a prehistoric lizard framing a heart was carved in the center over the main stone entrance to the room. The mystery deepened. Perhaps it was a clue to the timeline disturbance and paradox the Regent had spoken of. Jeff remembered the Regent's words, "You will know it if you find it," and they resonated in his mind.

The 20th-century toy box, adorned with a faded rabbit, the paperback child's book, and this leather-bound journal - the mystery deepened. But he couldn't confide in anyone. He wasn't ready to surrender this document to the Regents; he needed to read it to uncover its secrets.

The first image was an old black-and-white photo of a young girl standing in a rainstorm. It looked like it was from the 1950s or 60s. The book seemed to be a handwritten novel or maybe a young girl's journal. He gently turned the page. Could it be a photo album? It felt too contemporary. The odd black-and-white image held Jeff's gaze as he pondered.

# CHAPTER 3

# The Manuscript

Jeff gently turned the page of the mysterious, leather-bound book and began to read. The handwritten document appeared to be either a journal or the beginning of a novel. Cowboys and Indians seemed odd and out of place in this ancient cliff dwelling. However, adding the word 'alien' to the title piqued Jeff's interest.

**Cowboys, Indians and Aliens**

The year 1962 was an interesting time in my life. It was the last year of elementary school, and I was just about to start the sixth grade at Bassett Jr. High School in El Paso, Texas. In other states, it was often called middle school, a description that seems more accurate. It was the summer when the line between child and young adult became somewhat foggy. I felt too young for some things and too old for others.

Throughout my five years in elementary school, my best friends were boys. I had always been considered "one of the boys." It wasn't because I was a tomboy; it was merely an issue with my neighborhood. The only children to play with were boys; not a single girl was present in all the years I lived on Nations Avenue. I had long blonde hair, which my mother would put in two long braids. I was the Indian who would be chased, captured, and scalped. Don't ask me why they thought the cowboys were the ones who scalped people, but that was my lot in life.

I shared my fondest memories with a gang of seven boys. They were my bullies, my best friends, my companions. At some point, they must have realized I was a girl, but initially, I was just the smallest and the only Indian on the block. Rumor had it that we had Native American ancestry, but I looked more Scandinavian.

My long braided hair shone like the sun, and my upturned nose was turned up like a pixy dotted with freckles. I was a skinny, freckle-faced girl, but it seemed the Indian braids were all the boys needed for their game. My greatest advantage was my speed. I could run faster than any of them, so I spent most of my playtime running. You might have thought I would become a track star, but that didn't happen.

My neighborhood was a long straight street up the side of a mountain. All summer long, I would walk the trails up the mountain farther than my mother's whistle could reach. That was how I avoided the daily chase, capture, and scalping the neighborhood boys looked forward to.

They would stay close to the street, within earshot of their moms. But I was different, truly a maverick. I would test the limits of freedom and often venture a bit too far. Many times I could hear my mother's voice in my head telling me to come home.

"Don't take too long to get home, or your father will have to scout you out," she would say.

Both my parents were telepathic and keen sky watchers. My father was a scientist and teacher, and my mother was an English teacher and coach. Evenings at our house were never about sitting in front of the television set but about climbing a ladder to lie on the roof, watching the stars, constellations, and UFOs pass by. I often wondered if my parents were aliens, waiting for the mothership to buzz our house as a sign of greeting.

My fascination with everything extraterrestrial was a little over the top most of the time. Considering the life I had at home, I feel I can forgive myself for that. Conversations about science and the universe were everyday occurrences in our house.

Having a scientist, inventor, and science teacher for a father meant I was always hearing about the cosmos, the visitors, and the book my parents wanted to write about the people who lived on the planet Titan, now just an asteroid belt between Mars and Jupiter. They would spend hours discussing the Science Academy and the Great War in the Cosmos; my mother's best friend would keep notes as if she were helping write this book.

My father had a "Science Club" in the back workshop of our house. His brightest high school students would come once or twice a week. They would laugh, talk, and discuss the known world as well as the theoretical physics of space flight. The old-fashioned green blackboard was always filled with mathematical equations scribbled almost in circles.

As a teenager, I would watch from a distance, drawn more by a banquet of buff guys than by an emerging interest in science. The onset of hormones was clouding my scientific mind, or at least that's what I would tell myself.

It was in that workshop that my father and his students worked on a system to manipulate the lines of flux to create anti-gravity. Even though I was and am still fascinated by all things strange and other-worldly, I didn't participate in this high-level intellectual exercise.

Jeff stopped and shook his head. Almost out loud, he said, "Anti-gravity… using the lines of flux… and positive and negative energy flows… that wasn't developed until 2034. How could this high school science teacher have developed this power source and the world not know about it back in 1962?" There was something odd and paradoxical about this document; it couldn't be a work of fiction. The facts were accurate; it must be a diary or memoir. Jeff leaned back, sipped his now-cold coffee, and picked up the document again, hoping to unravel more paradoxical science.

I was a bit provocative for my age and thought the boys were cute. One Saturday afternoon, I put on my new bathing suit, a one-piece my father thought was cute. I rubbed down with plenty of suntan oil and walked past the workshop window. A model of a disc-shaped craft sat on the work table in the center of the room. The guys were so engrossed with this toy they didn't even notice me.

I walked up the side stairs to the roof and laid down my beach towel. The sun was a bit too hot, so I lay on my stomach. Suddenly, I felt a vibration. The building seemed to shake as if the roof would fly upwards. I jumped up, and I think I screamed, then I ran down the side stairs to the window. My eyes flew open so wide my eyebrows must have been stuck in my hairline. I was shocked to see that the toy model, the disc-shaped UFO, was hovering on the ceiling. All eyes were glued to the tiny craft when it suddenly dropped to the floor.

I rushed through the door. Everyone instantly looked at me as though I were a spy or something. "Hey, what was that? The roof almost came off. Are you guys building a flying saucer or what?" My father looked at me with those eyes that told me I was in trouble. He pointed to the door, and I backed out.

I heard Dad say, "Well, I guess it is back to the drawing board, guys." The students inside laughed and shook each other's hands. I think their experiment must have been a success, even though the workshop roof remained attached to the building. My father walked towards the door and closed it. I stood there looking at the closed door for what seemed like hours and finally just went inside and changed into my Capris. That was the only time the tiny model ever flew.

Now that I think about it, that was the last time the model was seen by anyone. Do you think my Dad took it outside that night, and it just flew up to the upper atmosphere? You know, I never asked him, nor did I ever discuss what I saw inside the workshop that day until now. I had recorded the event in my journal, the key-locked diary my mother had given me for my tenth birthday.

The last week of school seemed to drag on, and finally, school ended. The flight of the UFO model was the last meeting of the Science Club until next year, and I was sort of sorry to see it end. I so liked to daydream about flying off to space with one of those handsome high school boys.

Jeff placed his white linen gloves inside the leather-bound journal, slipping it into a pillowcase. He hid it under his cot for the night but would hide it out of sight in the morning, placing it under his pillow. After the first day visit by the Regents, Jeff didn't want to have them come and remove his find.

His mind was overwhelmed with thoughts of this young girl's diary, which read like a historical journal. He fixated on how she may have caused a ripple in time. He yawned and stretched. His mind fantasized about what could happen if a person took the path less traveled while the other continued forward. Was this potentially how a time paradox works? His mind traveled to all possibilities. Sleep was coming. In the dark recess of his mind, he thought about what his destiny would have looked like if he hadn't taken the risk of moving back to Arizona.

By bypassing the option to become a Professor at Harvard, life would have been altered, his children would not have been born, he would never have met his soul mate, he would never have had to suffer her loss, her untimely death, his unbearable sorrow. A tear rolled down his cheek, and he covered his face with his pillow. Tomorrow was a new day. He would put his past behind him, bring pleasant memories forward, and get a good night's sleep. The dig waited. The adventure had just begun.

The plans were to bypass the large main hall area of the cave-dwelling and venture deeper into the cavern. This was the area not ventured into back in 1909. The deadly smell and the profound darkness had prevented Professor Jordon from discovering what Jeff had surmised might be where the major Cibola city complexes began. Yesterday's passageways had continued to drop in elevation by over a mile. The inky black crevasse couldn't be that much deeper. Jeff hoped and wished to himself.

It had been proven by historians that the mythological accounts of the past had turned out, with research, to be historical accounts of ancient history. Cibola existed. It had to. He would concentrate on finding what the Regents funded him for. Cibola and the Egyptian connection, the Egyptian connection had already come to pass, both now and back in 1909. Why would the Regents bury the artifacts on the secret floor of the Smithsonian Institute's Museum? That forbidden place where strange artifacts are sent to die a death of 'out of sight, out of mind'? He hoped that would not be where his finds ended up.

# CHAPTER 4

## Can you dig it?

Jeff woke up and dressed in his khaki pants and shirt. He looked the part of an archaeologist. He opened the flap of his tent, took a last look at the hiding place of the journal, and stepped into the morning sun. He smelled fresh coffee brewing and what he thought to be pancakes.

"Professor, I decided we needed bulk this morning, so I have made blueberry pancakes for everyone. Come and join us," Sally, one of Jeff's brightest students, said, pulling her long auburn braided hair away from her face. Standing up and pressing her chest forward and shoulders back, she revealed her artistic figure.

Her face was beaming with pride as though the pancakes had been a masterpiece. Jeff didn't know how to politely tell her she was wasting her efforts on him - she was the same age as his daughter, for heaven's sake. She was trying to attract his attention, but he simply ignored her advancements. Nevertheless, she made a point of handing him a cup of coffee and a huge plate of pancakes.

The group sat together and planned the day. Jeff placed his half-eaten plate in the middle of the table. Sid looked at it and asked, "You done with that? You mind if I?"

"All yours, but we need to hurry," Jeff replied. Sid gobbled down the last of the pancakes and patted Sally on the rear as he walked over to the climbing ropes. She didn't seem to find offense. Perhaps it was because, with them, she was just one of the boys. As they gathered their packs and headed for the pulley to descend into the cave entrance, a hovercraft rose out of the canyon. Jeff's heart sank instantly. What now, he wondered.

The craft landed, and the same two Regents stepped out. One of them raised a hand as a hello, and the other seemed to be examining the pulley system. Jeff walked halfway toward them and waited. The taller Regent spoke first.

"Professor Ryan, the Smithsonian would like to procure the relics you have found this afternoon if that is agreeable with you. We need to move on this as the Environmental Protection Agency is causing such a media storm we fear the expedition may be cancelled."

"Those tree huggers are more hot air than action. You know that. I don't want the dig disturbed while my students and I are exploring. Any movement of statuary or any artifacts could have safeguards that could cause cave-ins. I have seen this happen more times than I can count. Your liability and moral responsibility for the University students should be paramount to the Board of Regents at the moment. After all, you are paying for the dig. That makes safety your concern. I will have to ask you not to proceed and, as I previously suggested, give us at least three months."

The first Regent shook his head and finally simply said, "Fine then. You have your three months."

The other Regent walked towards Jeff, looking suspicious and, frankly, fearful. "May I ask why you haven't requested the extended use of a hovercraft to access the cave entrance? It would seem to make your expedition more expedient."

"Why keep a pilot on extended hours just to do ten minutes worth of work? Besides, rappelling and using the pulley system is part of the experience for my students. Not all archaeological sites come with fully funded grant money. It is part of the learning curve for future historical relic retrievers," Jeff responded, smiling to himself as that phrase summed up what he thought of the Board of Regents interest in antiquities: things to be retrieved.

"That makes sense, Professor. It's no wonder your students always transition from the degree program to proficient professional archaeologists. It must be all those teachable moments. We will let you get to your dig. Good day, sir."

Jeff waved but did not respond. He acted as if he had forgotten some equipment and entered the van, looking out the side window, waiting for the hovercraft to leave the area. When they did so, Jeff jumped out of the van and yelled, "Burning daylight, children. Let's go."

The students arrived at the dig in record time. They had the rappelling and the use of the pulley system down to a science. That morning, they would all enter the main hall, something only Sid and Jeff had seen, and they would see one of the family living areas before venturing into the unknown. Jeff would be sure to guide them to the empty area and avoid the room which still contained the wooden toy box.

The map that the Professor was using ended after a long corridor from the larger hall area. There was nothing beyond that. It was possible that it was because the corridor was a dead end, although the previous day, they thought that even in the inky blackness, they could sense a deep cavern. If it were a dead end, they would divide up and venture to the long east and west corridors, which, according to Professor Jordon, had a cache of sarcophagi likely laden with gold.

That was unless the Smithsonian of 1909 had already been the grave robbers. Jeff knew that would be the preferred direction for the Regents because he was beginning to understand that their motives were not about history or the mystery of the Egyptian connection within the Grand Canyon but a gold rush to 'expand their holdings in the galaxy.' Although Earth was not functioning on a monetary system but a community contribution method, the galaxy traded in gold.

The map was deceiving as it wound downwards for miles. The crevasse turned out to be lightless corridors straight down into the earth. The students each filmed the walls whenever they saw the artwork, hieroglyphs, and even petroglyphs, which were indeed out of place. The two symbols should never appear on common walls, nor should the carbon dating machine claim them to be the same age. It was a paradox.

But, since the students hadn't noticed, he kept his thoughts to himself. The darkness increased exponentially as they descended deeper into the cavern. Jeff held his ultra-luminescent wand out in front of him. The contrast between the extreme darkness and the brilliant light made it hard to see directly in front of himself. He stepped cautiously. Sid had become his rudder during the descent.

Jeff rounded a sharp corner, and the light wand suddenly flashed on an enormous crystal cavern. Jeff began to lose his footing and looked down at what appeared to be an endless crevasse. Sid grabbed the Professor by his belt and yelled, "Someone help. I can't hold him. My hand is slipping."

Jeff felt his body lunge forward, and his feet began to slide on the tiny stone he was perched on. Another rock climber slipped between Sid's legs like a ball player sliding into home base, grabbing Jeff by the legs with both arms. The motion gave Jeff the push he needed to come off his stone entirely. His body lunged forward, head facing the now twinkling stalagmites, as his light wand fell.

The wand, in its seemingly endless fall, lit the darkness, and the shaped columns of crystal gleamed. He knew they could have impaled him. Panic set in as he realized nothing stood between him and death but his students. His life flashed before him, time stopped, and his mind went blank for a moment.

Then, he felt the full pressure of the ground and Sid's body weight on top of him. It brought his mind and soul back to the gravity of what had almost occurred. Jeff pushed Sid to the side and sat up, crawling towards the far wall. He placed his back against the wall and put his head between his knees. He didn't voluntarily let out a huge sigh, but it happened. After all, he was only human, not a superman.

Sally dashed over to the professor and pulled out her canteen. "Here, Professor, take a drink. I think you need it."

Jeff pulled the canteen to his lips and took a large gulp, swallowed and began to cough. "Sally, this is vodka. What the hell? That is not allowed at the dig, ever."

"Sorry," she responded, "I just brought it to celebrate a great find. It won't happen again."

"No, Sally, I'm sorry. Thank you. You were right… I needed it. Thank you, guys… for saving me. That is what teamwork looks like." Just as Jeff began to relax, he heard Sid let out a loud exclamation.

"Oh my God, you gotta see this. Man, you won't believe your eyes, but be very careful. There isn't any floor."

Jeff stood up and handed Sid his spare ultra-powerful light wand. Sid held his rock climbing buddy for stability and leaned out with his arm extended as far as he could safely go. The opening to the new entrance was about five feet across, which didn't leave a lot of room for everyone to peer into the darkness.

They divided up into groups of three. Jeff, Sally, and another student crowded around Sid and his buddy and peered into the darkness. The light wand at the bottom of the cavern and Sid's above provided a spectacular view of the area below. An audible sound of excited inhaling echoed into the cavern below.

"Crystal, crystal as far as the light will allow us to see," Jeff's voice echoed his excitement, as well as echoing down below like a call and response. "Stalactites and stalagmites of twinkling crystals. I think I see the edges of a pyramid in the distant light. Can you tell, Sid?"

"It kind of resembles the outer shape, maybe. Wow, what a find. But how are we gonna get down there?"

Jeff stepped back to let the other students get a bird's eye view and see if they concurred with the impression that the lost city of Cibola lay below, pyramid and all. It was a find that would bring more accolades than he was used to, but he hoped the Smithsonian would allow the media and historical data collectors the opportunity to archive the find. Each student took video logs, and the Regents would have full access to the find via telecom.

It was only 3:30, but it was more than a full day. Jeff gathered up the students, and they made their way back to the pulley. It was time to relax, ponder the find, and for Jeff to read more of the mysterious journal. He had decided to tell his students that he was researching an old document he brought with him from the Smithsonian.

It was a priceless leather-bound journal of the Lewis and Clark expedition. He wanted to see if, somewhere in this document, he may find mention of the cliff dwelling. He informed his students that this was for his hands and eyes only, as the Regents made him sign a waiver to take it from the museum. Jeff had figured it would give him more opportunity to read while outside his tent. This story gave him a logical reason to have this manuscript.

The evenings on the mesa were breathtaking, and the stars were plentiful. Before Jeff gathered the journal, he and the students brainstormed plans to descend into the crystal city. They decided that one student would use the rocket pack, which the Regent insisted on leaving with them, to inspect the stability of the sidewall into the cavern. If it was possible to use the magnetic pitons on the wall, depending on the iron content of the rock, a ladder could be created. It sounded like a plan, a dangerous one, but a plan.

The following day, Jeff would join his students in evaluating the feasibility of this form of travel. If, as the Regents were hoping, the walls were made of quartz crystal, it would figure that gold veins ran everywhere in the descending wall. If that were the case, the magnetic pitons would not hold. They needed iron and strong metal content to the rim, iron pyrite but not gold.

The students wandered off to celebrate in their own way, and Jeff hoped there wasn't too much vodka on the premises as he didn't need impaired climbers the following day. For now, he thought, to the journal. Jeff retired to the tent and retrieved the book from under his pillow. He opened it to the white gloves and his bookmark, put them on to prevent the acid from his fingers from damaging the fragile pages, and began reading.

**ET, spy or what?**

Summer was in full swing. I had just finished mowing the yard with the old push mower—a penance for violating the science club meeting the previous weekend. I mowed everything green, including all the weeds.

In my mind, anything that could look green in the desert deserved to be called the yard. I had decided to wear my new two-piece bathing suit, even though my father had thrown a fit the first time I put it on.

He had asked me how much I paid for it, and I had told him twelve dollars. He had said, "I could have given you an outfit like that for two dollars - two fifty-cent pieces and a dollar bill."

The suit really wasn't that skimpy, but Mom told me it was because my Dad had just noticed I was looking more like a woman than a little girl, and it scared him.

Jeff laughed to himself, remembering the moment he discovered his little girl had matured. No matter the timeline, the emotions of a father rang true. So, I wasn't really supposed to wear it anywhere near my Dad. Mom and Dad had gone across the street to play bridge, and I knew they would be gone for hours. I was supposed to keep an eye on my little brother while they were gone. He was easy to watch because he was always reading. Mom always called him her "egg head," her way of saying he was a brainiac.

I came inside to get a cold drink and bother my brother as much as possible. He was curled up in my dad's easy chair. I was about to interrupt him when I heard a knock on the door. I hesitated. Mom and Dad were not home, and we were not supposed to open the door to strangers. I couldn't see who was out there by looking out the living room window. As I stood at the window peering through the crack in the Venetian blinds, my brother just opened the door.

I dashed to the front door, knocking my brother into the wall. It really was an accident— I was protecting him. He left a nice imprint of his backside on the sheetrock. As I watched my brother slide off the wall like wet mud and then to the floor, I rushed to the now wide-open door. I saw three strangers crowded onto the front step. There was a tall, muscular man with a military-cropped haircut and two tall, large-boned German-type women with blonde hair looking at me and my prone brother. They appeared to be wearing some type of uniform, but there was no visible logo. The women almost reminded me of stewardesses.

"We aren't buying anything, and besides, my folks...well...you should just come back later," I gulped as I slammed the door in their faces. I felt really bad about being so rude, but what was I supposed to do? I was wearing that sailor-style two-piece that my father thought was too skimpy for the public, and I was not supposed to open the door to strangers. I was so mad at my brother. I just knew my folks had envisioned this event in their minds, and I would have to do some other terrible chore to make up for not taking better care of my brother.

As I turned around and helped my brother off the floor, I felt eyes on me. I turned, and there were those three weirdoes sitting in the living room as though I had let them in or something. I felt my heart stop and panic race through me.

I really don't have any memory of when my parents came home. I just know that suddenly we were all together in the living room with those three people. The tall man spoke and shattered the silence that was in my mind. "We have come to ascertain the reason your house is registering due north for more than one mile in radius in a three hundred and sixty-degree circumference."

Dad had discovered by accident that north was our house all around the neighborhood. He hadn't considered it an important enough anomaly to worry about, although it had rained on our backyard twice in the week that followed and nowhere else. He had deduced that while crossing the lines of flux, using the magnetic poles, he might have altered the ground under our house. We had talked about it but dismissed it as a side effect that would wear off soon.

"My students and I have been experimenting with gravity and magnetic energy. I think you may have some misinformation," my father began to say.

I knew that during this time, my family was guarded against rumors and misconceptions by those who did not know the family. It had only been eight years since the end of the McCarthy Era, and people were slow to come to terms with the fear of the communist threat.

Jeff paused for a moment as if to search his memory of any history concerning this person, McCarthy. He had been a history teacher when he first got his Bachelor of Science degree. It had been a long time, but still...it didn't seem to register.

Everyone suspected everyone. It was natural to be guarded, particularly if you were building a spaceship. Everyone in my community knew of the communist scientist who had come to live and work in the area. These gifted Nazi scientists were given citizenship so they would develop missiles and technology... methods of exploring outer space. The fact that Dad was building a model of a

UFO was still viewed with the same suspicions as those communists housed at Los Alamos.

At that moment, my mind went blank. There was fear of the government and the history of the McCarthy arrests. I was so traumatized by the event. They had come in without my knowing it... almost like magic. The time from seeing them in the house while my parents were not home to seeing them in the house with my parents seemed to be in the same moment. How could this be? I don't know if it was simply that I couldn't remember this memory or that it was somehow erased from my mind... but that was the last memory of the strangers.

The sun entered through the slit in the window shade at an angle, indicating that it was nearing sunset. It had been three in the afternoon when the strange people arrived at our doorstep. My mother was sitting on the sofa while Dad was in his chair. My brother and I were on the area rug. Books were spread out as though we had been reading. There were no weirdoes in sight. I looked at my parents, who appeared as puzzled as I was. We realized that more than two hours had simply vanished. The question of what happened to that missing time still haunts me. We never discussed this lost time. It seemed as though we chose to ignore the questions in our minds and merely continued with the rest of the summer.

We did, however, talk about the peculiar people who came to ask Dad about his Science Club and its experiments. Dad had concluded that we were either visited by secret government agents or Werner von Braun with two other scientists. He hypothesized that we were kidnapped, interrogated, and then given something to forget the event, or perhaps, we had been abducted by aliens. It was always entertaining to speculate, imagine, and ponder these mysteries. We lived outside the White Sands Missile Range and less than one hundred miles from Los Alamos, the secret community out in the desert that the government established to house the Nazi scientists after the Second World War.

Jeff remembered the history of the German scientists who created technology for space travel. However, he hadn't read any history about missiles or warhead-type weaponry. The world had been at peace for the last hundred and fifty years or so. The young lady was either writing from a history that never happened, creating a science fiction book featuring inexplicable acts of war, or perhaps these events had indeed occurred. Was there a rip in time? Had something or someone altered history? He was more compelled than ever to discover the truth behind this journal.

White Sands Monument, located outside of Los Cruces, New Mexico, is surrounded by missile ranges and secret underground laboratories. Just beyond this ocean of sand lies the city of Los Alamos, where the scientists resided.

Jeff lifted his gaze from the pages and stared thoughtfully at the tent roof. "Wow, I do remember studying German scientists in my sociology class at ASU. Van Braun was moved to Fort Bliss, Texas, a scientific compound, in 1945. I knew he worked for NASA. I thought he worked in the research and development complex at Dulce Base. I was taught that only space flight was being developed at both locations. Was there more secret information that the public had been misled about? Had the supposed missile ranges been redacted from public documents? Why would this child even think about them if they didn't exist?" Shaking his head, he resumed reading.

> The National Park area was an enthralling 275 square mile phenomenon created by pure white crystalline gypsum (hydrous calcium sulfate). These crystals grow and move at a rate of 30 feet per year, creating great waves of white sand, an ocean of sand. The snow-white dunes were popular among us.
>
> During the summer, we enjoyed driving to the park to go sand surfing. Though it wasn't a well-documented sport, it was more manageable than ocean surfing. We used car floor mats or waxed boards to glide down the dunes. The landings were hard, much like hitting cement. Unless one could loosen the top layer of sand, it was fast and painful but also fun.
>
> One Saturday afternoon, my friends and I arrived with an older brother who had agreed to our adventure. After hours of running up and down the dunes, we saw a triangular craft hovering in the distance. We wondered if it had come from the secret labs or if it was an alien craft.

Gathering our picnic supplies, we got into the guy's car and began to follow the craft down Old War Road 11. This road was off-limits to non-military personnel, but the older boy told us he and others used it for racing once a month. The craft seemed to wait for us to catch up. We followed for miles until the craft abruptly ascended and disappeared from sight. The boy driving braked hard when he saw military jeeps blocking the road. I was afraid - my parents would be upset that we hadn't taken Highway 70 straight home.

"You are on a restricted road," yelled the uniformed army sergeant. "You will follow us to the junction of the highway, and you will never use this road again. Do you understand?" He said to the other officer, "Those damn teenagers use this road to race. I'd like to confiscate their cars. Some people's children!" He raised his arm and signaled for us to follow him.

This off-limits road was constructed to transport missiles and missile launchers to and from Los Alamos and a secret underground lab at Fort Bliss in El Paso, Texas. The underpasses of the regular highway were not tall enough to accommodate missiles. Rumors had it that this old highway was used by the military and the scientists to avoid detection by non-military personnel. My mother dismissed it as a waste of taxpayers' money and an abandoned highway. However, it was quite evident that they were still monitoring this stretch of road.

"We just got lost," the driver began to explain, but the army man was already heading towards his jeep. He gestured for us to follow him, and we trailed behind the convoy until we reached the main road and sped home. None of us mentioned the UFO or War Road 11 again. There was a real fear of the military police in our town, our neighborhood. This fear was a lingering effect of the McCarthy era, a time when everyone was suspecting everyone else. It didn't end until 1954, and in some places, the suspicion lingered. Our area was one of those steeped in secrets, making it natural for everyone to be on their guard.

Again, Jeff stopped to consider the information. There were too many actual dates and facts to be a work of fiction. He questioned, had this document recorded an alternate reality, one that no longer existed in this timeline? How could this journal have appeared inside a wooden toy box in an ancient dwelling? He pondered the Regent's last words to him. "You'll know it if you find it." Perhaps he had, but what would the knowledge of a history that didn't exist in his history books mean to the Smithsonian? Why would this information be of such concern? It was just weird.

Late one August night, my whole family was on the roof looking at the stars. Suddenly, a comet or meteor became visible. The comet came even closer, and I could make out the triangular shape, the same shape I remembered from our White Sands adventure. The craft whizzed by the house, making a whistling sound. My father sat up and said, "They are either landing or crashing."

I was afraid to tell my father about the craft I had seen in the desert because my friends and I had ventured off the reservation onto restricted government property. I didn't want to bring trouble upon myself. Mom and Dad discussed the sighting and waited for news of some astronomical event or newspaper reports of a UFO sighting, but none came.

I wondered, was this ship landing? Would we ever know what passed by our house? Were we in danger? After all, it had been less than a month since we had been abducted or visited by government scientists who drugged us, so we couldn't remember the event. Perhaps it was just a shooting star. It couldn't have been the same craft I had seen just weeks before. I didn't want to worry about the possibilities. As a world-famous sleuth and master of avoidance, I preferred to deny and ignore rather than face an uncomfortable reality.

Summer vacation had ended, taking with it all the unusual events and happenings that consumed my time and imagination. The first day of school finally arrived. After a summer of adventure, I was more than ready to return to school. After all, I was going to be in junior high school. Most of the country called this middle school, which was probably a more accurate description of this odd life stage. Excitement was the rule of the day. I was worried about getting lost. This was the year I would have to learn to travel between classes and manage multiple teachers, not just the one teacher and one coach I had become accustomed to. I made sure my hair looked fashionable. I had my mother cut my hair into the shoulder-length style that I saw on American Bandstand. I wore a great sweater set and a matching skirt. I felt really grown up.

Although I knew many friends from my neighborhood and elementary school, they all seemed to be living in their own worlds and took little notice of me. I kind of missed the days of playing cowboys and Indians. I felt like I was in the Twilight Zone, not a kid, not an adult, just a lonely creature in transition.

Weeks went by, and I began to feel lonely. One afternoon, I was rounding the corner to my science class when I bumped into a tall, strangely beautiful boy in the hall. He had hair the color of spun gold, which was a little longer than the norm. His skin was soft cream, and his eyes were the brightest blue I had ever seen. I stood staring at him. He looked at the ground, and I suddenly noticed his unusually large head, which looked a bit like the bust of Queen Nefertiti. My stare had embarrassed him.

Jeff thought it odd. There were no records of the Saurians visiting this part of the Southwest during that time in history. The first official contact with the Galactic Federation wasn't until after 1980, and the World Alliance was formed around 1981. Strange, he thought, as he continued to read this mysterious account of a history that didn't match his history books.

"Sorry, I didn't mean to stare, but you have the most beautiful eyes," I said, attempting to make him feel less self-conscious. He straightened up, and I thought to myself, he should join the basketball team.

"Hello," he replied, "I'm an exchange student here. Maybe it was my fault. I'm not too sure which classroom I should be in. I know you didn't mean harm or ill will towards me with your gaze." His eyes were gentle and understanding, and I felt I could just melt. He seemed out of place, and his deep, warm voice seemed to massage my body.

"So, where do you come from?" I asked, curious about his appearance, the unusually elongated skull, his odd voice, and everything else.

"I come from next door," he whispered, quickly walking into a classroom.

I couldn't get him out of my mind. Who was that kid, that oddly handsome boy? I thought he said he came from next door. I wasn't aware of anyone who had moved out all summer long. I knew everyone up and down the block, even the cat lady up the street. I didn't remember anyone new to the area. Understand that I am a need-to-know girl, and as they say, "Enquiring minds need to know." Although I never admitted to being wrong, I must have been mistaken. I decided to investigate. Maybe he had said Luxor or Dexter. But an exchange student, I wondered.

Weeks went by without finding that odd boy I met in the hall. I finished my investigation, which meant that I had bothered everyone in the neighborhood with questions, and found that there really wasn't anyone living next door who didn't belong there. No one admitted to ever seeing or talking to this tall, golden-haired kid. Perhaps I had imagined him in some weird daydream.

Regular school life returned, and the boys from the neighborhood finally figured out that although I no longer sported the long blonde Indian braids, I was still me. The silly games of chase and capture continued every afternoon. I no longer got "scalped" but became the spy from out of the cold, whatever that was. They became the FBI, and I was on the most wanted list.

Again, I took to the trails up the mountainside, investigating the caves and quarries that had been abandoned years before I was born. I took rock samples and quartz crystals from the caves and imagined myself to be an archaeologist looking for the lost city of the ancients.

"She's a girl after my own heart…could have been part of my family," Jeff thought to himself. "Early archaeologist by nature." Jeff was beginning to really like this young lady and her optimistic view of the odd and weird. There was a feeling they were somehow connected…that was random…this journal does make one's mind travel to strange places, he thought.

One late afternoon, I was reaching for a shiny rock that looked like foil. I picked up the rock, which was flat and layered, and thought of my father, who was originally a geologist before becoming a science teacher. He would know what this odd rock was. I slipped it into my pocket, and it felt hot, almost too hot. Was it radioactive, I thought? I mindlessly turned toward home. I was in for a big shock. You will never guess what was on this desert trail. It wasn't a rattlesnake or a bobcat but something that caused my heart to jump nonetheless. It was that boy, that odd and handsome boy, just standing in my path towards the house. It was going on sunset, and I had already ignored the voice in my head, my mother, insisting I come home. Why was he here? Where had he come from … what should I say to him? He didn't live in the area, and I wondered why he was suddenly on this path…my private sanctuary…my mountain.

The sun was to his back, and his golden hair shone with the light of the sunset, causing a halo effect to appear around his head. His white skin picked up the golden light that just dusted his cheeks with magenta. He was almost too beautiful to look at. Funny how you just know that your face is doing something stupid without your permission. My mind was a muddle of fear and excitement…wonder and dread. I knew when I questioned him.

"You told me you came from next door… but…but... I know you don't live there. Where did you come from, and why are you standing on my mountain?" I demanded. My face was blushed red with a combination of fear and excitement…and yes… worry that I would be late for dinner, making my parents unhappy. I had already heard my mother's signature whistle echoing up the trail twenty minutes ago and knew I was short on return time.

"Look, my name is Seth. My people tell me that they know your people and that it would be fine to tell you the truth. I do come from next door. I come from the Tau Ceti system and the planet your scientists call Tau CetiF. We call it Mira. Tau Ceti is the solar system right next door to yours. We are 11 light years away by your scientists' estimation. So I indeed come from next door." Seth spoke rapidly and in a hushed voice as if the cacti would overhear the conversation. "I consider myself to be an exchange student because one of the children from the hidden city was sent to take my place while I was here. Oh, and for your second question, why am I standing on YOUR Mountain? I am standing on what you call YOUR Mountain because this is where MY portal is to go home for the night."

Jeff stopped reading for a moment—Tau Ceti? The enlarged skull is of Egyptian, Sumerian ancestry, more reptilian. Tau Cetians are humanoid…at least, they are where I come from. Jeff shook his head again. This was really mind-blowing in so many ways.

I know I must have looked stupid, but I stood my ground and finally put my hands on my hips and flatly said, "Fine then, I have to go home, it- it is getting late,

Goodbye!" I always had to have the last word. My mind flashed ...portal...solar system next door...my people know your people...what.

Seth pushed past me on the trail, and as his body brushed against mine, he said softly, "My portal will close soon, so I also have to hurry, but I see now! How our destinies will and must cross."

I practically flew home. What was he talking about? Will our destinies cross? Was he psychic or what? I wanted to tell the story to my father because, of all the people in the world, he would be the most open-minded and would give me the advice I needed just in case Seth and I bumped into each other again.

What was this secret information? Did my parents know something they kept hidden from me...really? I'm grown up. I did need to know.

Jeff placed his gloves inside the pages and gently closed the book. This young girl must have been the catalyst for the time paradox the Regents hinted about. That type of knowledge, portals, time travel. Was this information science fiction, or was it?

At any rate, at the time frame, this journal was written. Some of her information was genuinely scientific, yet not widely known to the general public. Jeff hadn't even learned of some of the information the girl had written in her journal until very recently. The mystery of the paradox and what might have been altered by this child became an obsession.

Jeff would not let this document out of his possession until he could find the thread that unraveled time as this girl knew it…as he knew it.

# CHAPTER 5

# JACOB'S LADDER

Jeff placed the journal in the pillowcase and shoved it under his pillow. Dressed for the day, he was ready to take on what seemed to be an almost inaccessible cavern.

Quartz crystals would shatter if hit with metal pitons. Unless there were heavy metals and a large stash of iron in the sheer mile drop into the cavern base, there would probably be no way to access it, at least not without the Regents sending a small shuttle craft that could be flown into the long corridors to the cavern.

Jeff didn't want to bring them in on this find because they would instantly use the equipment to strip the artifacts out from under the research team, his brightest archeology students.

The longer he read the journal, the more secretive he felt he needed to be with them. After all, it appeared that something very odd had happened to history. It seemed to have been changed. Nonsense, he thought to himself. This girl had indeed gotten into his subconscious mind, his creative thinking mind.

His students were gathered at the table in the center of the camp, laughing and drinking coffee. Sally bounced over to Jeff, eagerly handing him a cup of coffee. I'll have to do something to dissuade this kid, the professor thought. University student or not, she was his student, and he was thirty years her senior and her instructor. It wasn't ethical or legal, or wanted, Jeff reflected, accepting the coffee and sitting next to Sid.

"Today is probably the most exciting but the most dangerous adventure we have taken to date," Jeff began. "Jacob is not only an expert rock climber, an advanced archeologist but holds a Master in Geology. Jacob, are you going for any particular field, or are you a career student? Never mind that. Anyway, because of Jacob's vast knowledge of everything educational, he will be the one to use the rocket pack tomorrow. He will sample the elements in the rock face and estimate what method we will use to descend into the cavern."

Jeff cleaned his coffee cup and stood up. "Oh, bring food for the evening and something to sleep in. We may be spending the night at the dig. Once we actually get down, returning to the opening may be even more of a challenge."

Jeff packed up NASA off-World Food packs for more than two weeks in his pack. He wasn't all that sure the climb out would be feasible. He packed the dry H2O packets and hoped for any drops of water coming off the stalactites to create potable water. He believed that one of the side tunnels that seemed to have radical declines might lead to the main pyramid. He surmised that the occupants from above would descend to the city below for worship or guidance.

It was the nature of the godheads and the common people. He so wanted to bring the journal along, but weight and space were critical for the success of the mission at hand.

Jacob flung the rocket pack over his shoulders and secured the straps tightly. He powered up and jumped over the edge, powering down to the cave entrance.

This left Sid to maneuver the side ropes on the pulley alone on the left side. Today of all days, the wind was blowing at 18 mph. Sally and Frank held the support on the right. The seven students took longer than normal to reach the cave, but once everyone was gathered, they proceeded down the mile of the corridor to the cavern entrance.

Jeff shuddered when his mind wandered to the fall he had taken the day before. A cold chill ran down his back.

The ultra-luminescent wand that the Professor dropped the day before shone brightly. Jacob lit two more and flung them widely across the vast opening. They floated more than dropped. It was odd, but possibly there was air flow from the outside. The wind was now howling through the upper corridors, making the ambiance more than creepy.

"Should I go, Professor?" Jacob asked.

"Whenever you have the stomach to make the leap, go for it. I suggest you venture out first so that you can take a good look at the rock face. If you would, send us the images as you go. Then take samples where you think the pitons should be fastened, and hopefully, we are on our way. Good luck, son," Jeff responded, holding his breath when Jacob fired the rocket pack and leapt over the edge. Jeff cringed.

The video Jacob sent showed large veins of crystal and gold but large sections of what looked like granite. Jeff called out, "Take samples in the dark rock to your right, and hopefully, we find enough magnetite or lodestone to make the magnetic pitons work. If nothing else, the granite-looking stone will probably be strong enough to hold the weight of our heaviest-laden explorer." Jeff laughed to himself, it was probably him.

While Jacob took rock samples, Jeff took the opportunity to check the weather station app on his eyeglass monitor. The weather for the Grand Canyon projected out in front of Jeff, showing the arrival of a haboob. The sand wall extended upward to fifteen thousand feet. All aircraft were warned away from the area from Sedona to the east, Phoenix to the south, and the Grand Canyon to the west. Jeff was really glad they made the decision to stay the night inside the cave. There wouldn't be any safe way to ascend the outer wall with pulleys or rope climbing gear. The wind was simply treacherous.

Jacob reported the news that the pitons could be set in the lodestone vein on the farthest side left of the opening. It meant that every student would have to swing out on rappelling ropes. Any mistake on the part of the inexperienced student would result in a fall of over a thousand feet.

"Okay, here it is," Jeff announced. "If you are concerned for your safety, which I would not blame you, you can bivouac up here. Sally and Sid can rig a safety harness, and Jacob will be at the first level to guide you to the double ropes he is currently attaching to each of the pitons. We will have to trust each other as we have to change pitons and ropes ten times before we hit the deck. The principle behind rock climbing and descent is to use the climbing harness and the safety belays, and don't let the double rope end!

"I will go first, I have the heaviest pack, and if there are any unstable holds, I'll be the first to find out. Sally will go next, followed by Tom, then Sid, June, and Frank. That will give us our four experienced climbers: Jacob, Sally, Sid, and Frank, helping your dear old Professor, June, and Tom. Four of you have decided not to take the risk. This is fine. I want you to remain here at the opening and keep the records of what we record below. You will create the report for the Regents."

"We should return to your level by midday tomorrow if all goes as planned," Jeff said, grabbing the rope which had been attached by a heavy eyelet piton to the solid granite floor. His hundred-pound pack caused his backward step off the opening to instantly pull him down. He slid, the ropes whistling through the eyelet piton.

His first emotion was sheer terror. Jeff finally caught his body weight using the harness and anchored himself, ultimately finding his feet against the wall. He began walking down the sheer face of the cavern, repositioning himself at the first stop.

Sally rappelled down to the first piton. Her body slammed against Jeff's, and she rolled her body around him and laid a big kiss on his cheek. Before Jeff could complain or comment, Sally rappelled down to the second piton's triple knot. The Professor followed until they both landed on the deck.

"Sally, you were to remain at the first piton to help Tom down. He is an inexperienced climber. Look, I am flattered by your attention, but I have to inform you that it is strictly against the University rules, and furthermore, you are the same age as my daughter. You're not my type, not age appropriate." Before Sally had an opportunity to make her case, Jeff called up the remaining students.

"Sid, it is now your responsibility to help Tom down to the second piton. Return and then help June with the first one. I will have Frank follow her down and take over at the second piton to assure everyone arrives safe and sound." Sid climbed back up to the first piton to help Tom down and make sure he followed proper protocol at the second piton transfer point.

June was the next to rappel down in tandem with Frank. She had a sudden panic attack as she attempted to take the first step off the ledge. Sid ascended to the deck, leaving Frank with June, the untrained climber, to the second piton transfer position. June seemed to have conquered her fear when she neglected the one rule, the most important one - don't let the second rope run out on the way towards the third piton. She was still six hundred feet off the ground. A scream echoed through the lower canyon and ricocheted back to the upper canyon, continuing for what seemed like minutes. She was falling at a great rate of speed.

Frank yelled for help. Jacob fired up the rocket pack and met her fifty feet off the ground. Her body bowed in his arms as the weight of the two of them brought them to less than three feet from the ground. Jacob released her limp body as Sid grabbed her, placing her on the crystalline gravel deck. Jeff rushed over and checked her health with his data card.

"She is okay. She just passed out." Jeff called out. His voice echoed through the vast citadel and upwards to the cave above.

"Dear God, I almost lost her. This might have been a very bad decision," he thought as he called up to the remaining four students.

He repeated his instruction to remain where they were. It was for their safety but also to preserve his secret find - the toy box. In his mind, he worried about his untrained students returning to the cave edge above. After this event, Jeff wondered if the rocket pack was strong enough to hold the weight of two. It was designed to hold two hundred fifty pounds and not much more. Jacob weighed about one sixty, and Tom was almost two hundred pounds. Maybe there would be a stairway upward to the dwellings above. He assumed that might have been the case. The occupants would have come down to pay homage to the gods at this pyramid if tradition and history were accurate at any rate.

The citadel was more than awe-inspiring. It was a sight that none of them had ever encountered before, not even the Professor, who had found the ancient city of Lumeria after the Pacific Ocean had receded, giving a glimpse of ancient ruins off of the Washington coastline.

Video recorders were humming in the stillness of the dank air. The wind was whistling from some unseen crevasses within the outer walls of the Grand Canyon's South Rim. Haboobs, although currently affecting the desert outside, must have impacted this deep citadel for centuries. The low howling added to the tension not just of the students but of the Professor. It was fodder for the imagination. Were there creatures within this citadel, unseen dangers?

The Professor had the students wait for him to make a sweep of the immediate area for danger from animals, creatures of the dark, pitfalls, or booby traps. He also wanted to check for any potential exit strategies. He carefully made his way towards the pyramid, which appeared to be about three thousand feet in front of him.

He walked towards the pyramid, which seemed to be constructed completely of quartz crystal, but as in those nightmares of running, the structure looked to be ever moving farther away as he walked. When he turned to see his students, he noticed that he had indeed traveled much farther than he had planned. He had been recording everything as he walked, and his students had been monitoring his progress. Jeff called the students and informed them that they could follow his lead. He would wait for them to catch up. Nothing dangerous seemed to be in his path on the way toward this odd crystalline structure.

The students met the professor, and they continued toward the pyramid. The size of this citadel deep within the earth seemed to be ever-growing in size.

The stalactites were so far above their heads that Jeff worried he wouldn't be able to find any droplets of water to create potable water in each student's canteen. As they continued on, the wind howled, and the air became damp.

The smell of fresh earth was palpable, like the coming of torrential rain. Soon there was the sound of rushing water, like a waterfall. It became deafening as the sound echoed ever louder as they approached the pyramid. The ground suddenly took a sharp elevation drop, and Jeff almost lost his balance. The combination of quartz, gold veins and what looked like mica was moist, making the ground slick like ice. Jeff practically skied down the incline toward the bottom of the cavern.

His body rolled as he hit the edge of what looked like a step. Landing on his back, he found himself looking upward at the cavern ceiling. He noticed a ray of light emanating from the side of the cavern.

The window to the outer wall, he surmised. No wonder the wind outside was so obvious. He inspected his surroundings for any sign of relics or a hint as to the origins of this odd structure, so out of place in the desert southwest.

As his eyes adjusted to the ever-increasing darkness, he noticed what appeared to be a small pond fed by a rushing stream of water. The more he stared at the water feature, the more he realized it wasn't a natural waterfall or watershed. It was obviously man-made. It was a great fountain, but where was the water source? As he scanned the structures that clustered around the great pyramid, the faces of his students now blocked his view. He hadn't realized he was still lying prone on the ground.

"God, Professor, are you hurt?" one student asked. "Man, that was a fall and a half. Let me check your head. There is a step right at the base of your neck. You might be injured." Sid ran his hand under Jeff's head, and Sally touched his brow as if to check if he had a fever. Jeff jumped up and brushed his shirt.

"I'm fine. The view from down here was kind of amazing. Really, you should try it." As he brushed off his accident, he realized that although Cibola was thought to be the city of solid gold the Spaniards searched for, this city was made of crystal. This made it one of the most potentially active power plants rumored to be within the inner earth. It was a myth that the earth's rotation was driven by quartz engines. Scientists had discounted this as nonsense, a science fiction fantasy of ages past, but here it was. The data recorder could detect the vibration coming off the structure, and it contained measurable power.

As he checked on the safety of all his students, he noticed Sid and Sally lying on the ground, their arms crossed as they examined the ceiling and the fountain of water which sprayed their faces. Sally brushed the moisture off her face and jumped up.

The rest of the students finally made their way down the slippery decline. They began moving about, recording everything of interest. Jeff instructed them to look for a staircase somewhere near the walls of the outer cavern. There must be a way to the surface, he thought. Meanwhile, he continued to look for a way inside the pyramid. Could it contain the full library, the secret history of the ages? He was on a quest.

He climbed stairs up to the side of the crystal pyramid. Each step seemed to grow in incline as he mounted the side. Soon he walked around toward the dark side of the structure. Even with the strong ultra-luminescent wand, it was difficult to make out the edges of the structure.

As he moved to the sharp edge, his foot began to slip. He grabbed something outstretched from the sharp edge. As he attempted to steady himself, the handle of stone moved under his weight, and the wall opened to a passageway. He almost fell, grabbing the outcropping with a vice grip. Finally, after righting himself and getting his breath back, he grabbed the light wand. He pushed his arm inside the opening as far as he could reach. The light exploded into prisms. He blinked his eyes and then looked again.

There it was - shelves of crystal jars, beakers of what appeared to be algae, and clear quartz skulls lined the upper shelf. Below were highly colored hieroglyphics telling the story of the structure, the people who lived within the city, and the tragedy that ended this age of humanity.

As he began translating the hieroglyphics, he read something that was completely wrong. The end of the age of humanity? This translation spoke of a cosmic explosion and the death of the savior, which altered two parallel universes. What explosion in the cosmos, and when did this history happen? His mind was tumbling as he stepped back and sat on the edge of the step he arrived on. That can't be correct, he thought to himself. Jeff had recorded the images but had not sent them in streaming mode. He would have to spend more time really analyzing the translations. He could have misread them. It would remain his secret, at least for now.

As he stood, the wall opening closed. He was glad he hadn't stepped inside; he might have been trapped within the pyramid. His communicator beeped, and he almost lost his balance again. He'd have to stop being so reactionary.

"Professor, we found a stairway. We will wait for you. It's under the waterfall. There's a tunnel that moves upward towards a spiral-looking stairway. It may be our way out," a voice came through the communicator. It sounded like Frank's—slightly high-pitched, nerdy characteristic, the sound of a slightly spoiled only child. His voice had a tonal quality that always annoyed Jeff.

"I'll be there in a few minutes. I'm on the other side," Jeff responded. He made his way carefully down the damp steps of the pyramid. He wanted to adventure longer, but it was really a difficult day. According to his data card, it was already seven p.m. He was getting tired and hungry. He was prepared with all the supplies other than the ones he left above to remain for the night. If they could all be on top, it would be a relief.

Jeff joined his students at the entrance of the tunnel, which appeared to aim in the right direction. Sid had already sent a drone to see if it actually ended on the main cavern floor or went off to yet another cavern. The results were streaming on the data collectors. They all read the results and took a vote. Stay and spend the night, ascend the same way they had descended, or take a risk and walk up the uncharted stairway?

The group decided on the risk-taking adventure. Jeff pulled out the supplies and gave the students the NASA gourmet dinners. It was nourishing, not tasty, but it would provide the energy needed to continue on. The water tested ultra-pure, so the students filled their canteens with water. Jeff led off and walked up the moist ramp towards the stairway. He hoped he prayed that it didn't just continue off to the unknown. The drone's report was sketchy at best. The darkness was profound, even with the best equipment. It was like entering a black hole - soundless, lightless, a life-sucking fear was in everyone's mind.

The climbing stairway ventured in a spiral direction. The light was too dim to see if the walls contained art or hieroglyphics. There was silence. Not even the sound of footfall could be heard. If one could not see or feel the person in front, it was like being in a sensory deprivation chamber. All sound seemed to be absorbed by the very earth that surrounded the students, the professor. It was unnerving.

After what seemed like an endless climb, Jeff rounded the last spiral, and an opening appeared. The light was less dense than it had been within the stairway. His ultra-luminescent wand picked up what looked like the edge of another tunnel. He hoped it was on the surface of the original cavern, but it wasn't part of Professor Jordon's 1906 map of the arms of tunnels the students had investigated. Instead, it was another tunnel. He checked his GPS. It appeared they were in a section just above the original map.

"How can we find our way back to the waiting students?" Jeff pondered.

He moved towards the farthest side of the tunnel and noticed a petroglyph with an image of a family—a woman, a tall man who looked almost Egyptian, a young girl, and two toddlers of the same height. The image of the woman appeared to be pointing toward the ground. It was odd, he thought. He looked down, knelt on the ground, and brushed the sand away from the floor. There was what looked like a hatch.

He found a finger hole and pulled upward. As the dust cascaded down the hole, the light revealed a stepping stone-type ladder that tracked down an exterior wall of the room below. As he aimed his wand toward the floor of the room below, he saw what looked like the edges of a wooden chest. A strange feeling arose within him—she seemed to be talking to me, he thought. However, he dismissed it as silly nonsense and cleared his mind.

"We're back, guys," Jeff yelled to the students following behind him. He wondered how he could get them to just follow him out the door into the main dining hall without noticing an artifact seemingly undisturbed in a room they would understand to be the one he inspected days before.

"Maybe just a bit of misdirection," he thought. "There is a hatch down to one of the rooms we found the first day. We can have the four students just meet us in the main hall."

"Aren't we going to go back to camp?" June whined. Jeff could tell by the tone of her voice that she'd had enough excitement for a lifetime.

"No, it is past nine o'clock, pitch black outside by now. I think we should just bivouac here for the night. I haven't decided what we will do in the morning. I'll have to sleep on it." Jeff got on the communication device and called the four students who were waiting for them at the entrance to the crystal city below.

"We are back up. We found a stairway, not great, but our way back at any rate. You can come and join us in the main dining hall. We will talk about the day and the finds and decide what to do from here."

A shy voice came over the communication device. Jeff wasn't sure who it was, but the message was plain enough. They had already decided to make camp. They were relaxing in their sleeping bags, and several of them were already sound asleep, according to the voice at the other end.

Jeff was glad he hadn't needed to call for help from the Regents in case of an emergency. It was a good thing this wasn't a graded exercise because they would have received a failing grade. They were obviously sleeping on the job.

The adventurous students made camp, rolling a light wand into a ball in the middle of the room as if they had a campfire. They talked and laughed, discussing the day. Jeff wondered if Sally had snuck more vodka into her pack to celebrate a great find, as she would say. It didn't matter anyway; they all deserved to relax.

What a day! Jeff put his sleeping bag in front of the entrance to the room, which contained the toy chest. They would have to walk over to him in order to investigate.

He fell sound asleep, dreaming about his finds—the crystal skulls, the hieroglyphs, and the petroglyph above. Was that the image of the young girl who wrote the journal? His mind flashed on a strange notion. The woman, not the young girl, pointed to the floor. Was the journal the mother's? It was odd, he thought, but he hadn't even begun to read more than a few chapters in this life journal. Who knew who wrote this? But how could this image direct him to the hatch? Was it destiny or just an odd coincidence?

Without that message, he would have been forced to walk back to the cavern below, through the soundless black hole of a staircase, potentially putting his students in harm's way when ascending the impassable wall. After all, June almost lost her life descending that mile-long drop. Had this message been left for him to find? It felt as if this female was somehow connected to him. This was the impression he had when he first began reading the journal. He slipped deeper into sleep.

His mind formed the image of his grandmother. She smiled at him and shook her head as if to say, "You're right… the message was left for you."

He woke with a start. Could it be true? No, he was just dreaming. Such silly thoughts were brought on by that journal, the toy chest, and his worry... his exhaustion.

"Sleep, sleep, please... just rest your mind," Jeff thought as he rolled over and put his back to the entrance. He closed his eyes tightly, listening to the students' laughter. They hadn't seemed to be concerned with their mortality.

"Ah, to be that young again," he thought.

The laughter slowly softened until sleep finally took Jeff back to another world. He dreamed of this girl, her family, and his connection to it all—and the crystal pyramid below. As he began dreaming, the story of their adventures made a movie in his mind. To dream, to sleep, and yet to dream again of a different reality and parallel universes, mysteries, time, and time unknown—he slipped into the silence of deep sleep.

Jeff awakened to the darkness of the cavern. It was impossible to tell morning from night, but Jeff's bladder knew it was morning. As his body stirred, his mind returned to the now. His watch read seven a.m. All his students were still sound asleep. He would use this time to investigate an impression he had while he slept.

They hadn't entered the east side tunnel number one—the very tunnel they decided not to investigate due to the petroglyphs seen in the west side tunnel. Could it be that this was the path taken by the early occupants to access the citadel below? He decided to investigate. He stepped carefully over his sleeping students as he moved back to the main cave entrance.

The sand had created dunes against the main wall at the opening and into both tunnel number one's east and west corridors. It looked as though no human had walked this path for eons. The wind had subsided, leaving the prevailing wind to blow the sand gently upward like miniature dust devils just inside the corridor.

He aimed down the east corridor, examining the glyphs as he went. They retold the story of the Hopi myth—*the One Heart* and *the Two Heart* who fought over ideology. The Hopi spoke of people's travels by boat across the great waters, following a guiding star to the canyon. Here, the Anunnaki ushered them into the inner Earth city to protect them from the great flood.

However, the find seemed to challenge this myth with reports of another earth-ending event, one not recorded in his history. This unknown event was inscribed in cuneiform—an anomaly at best, he thought. The great flood was a part of recorded history. He was beginning to understand that.

Jeff continued down the long corridor to a sharp corner and then a dead end. Stones had been piled, reminiscent of the work of ancient masons walling up the entrance of the corridor. Jeff had seen similar structures blocking the entrance to Egyptian tombs in Giza.

This area was directly above the Toy Box—the room that ended at the black hole stairway to the crystal citadel. This entrance to the stairway was impassable. He had hoped to guide his students through this corridor and down the stairway to what he assumed to be Cibola. His only other option seemed to be the room with the twentieth-century toy box. His find was still not safe.

Jeff returned to the main hall, where the students were just waking up. The four from the cavern entrance had finally awoken as June was preparing a pot of coffee. Sid looked up and yawned, "What's up, Prof?"

"Just checking for easy access to the city below," Jeff responded. "We'll still have to use those stone wall steps if we wish to return. I think we should go back to the main camp this morning and make new plans. The Regents will have had time to look at the streamed video by now and will more than likely already be waiting for us there. What do you all think?"

June practically jumped for joy. "That's a great idea!"

Jeff knew that when she returned to the main camp, she would hitch a ride with the Regents and go home. She was bright, or she wouldn't have been chosen, but she lacked a sense of adventure.

Jeff decided she would probably end up being a literary archaeologist, not a field worker. When this dig came to an end, he would recommend her to the Smithsonian Research Department.

Sid piped in, "Can't wait to get back there. What an incredible find. Was this where the beginning of the seven cities of Cibola starts?"

"Not really sure. Some of the cities have been found in Peru, one under the Giza plateau. What makes this find the rarest is that it is in North America. It had never been theorized that Cibola would be found here. Native American lore has hinted at its existence for a millennium. Even the early conquistadors traveled through Arizona searching for the mythological cities of gold, seeming to hint at Cibola's location. The very names of the mountains are Egyptian, like the Temple of Isis, my next dig—hopefully. I am confident the Regents will agree that this is part of the Cibola inner empire.

Sally ran over with a cup of coffee for Jeff. He took the cup and gave her a look that informed her that she should stop. She raised her shoulder as if to say, *can't blame a girl for trying.* As they sat with their morning coffee and the not-so-tasty NASA breakfast, Jeff heard the sound of transport in the distance. The sound amplified, followed by the voices of men approaching the group.

"The Regents. Look, let me do all the talking. Don't volunteer any information. Do you understand?" Jeff looked almost panicked as he searched his students' faces for the weak link.

A tall Regent approached. He wasn't wearing the typical tunic but more of a military-style jumpsuit. The uniform badge read "Smithsonian Investigative Bureau."

Jeff had never heard of this branch of the Smithsonian. It sounded like a policing board, but why were they there?

"You are to vacate the premises now," the officer stated in an official tone.

Jeff looked puzzled. "Why are we to leave? We have three months to finish this expedition."

"Not anymore. I have been instructed to remove you and your students so the Relic Retrieval team can remove the antiquities. We will collect any mapping you have accomplished to facilitate the removal of relics."

"Has the storm subsided? I don't want to endanger my students by exiting this cave in the wind."

"We will be transporting you and your students to the surface by way of the shuttlecraft. You will be able to gather all your notes and reports and leave the canyon by tomorrow. Is that understood?" The Regent reported in a staunch, authoritative manner.

"I suppose we have no choice in the matter. Kids, gather your stuff and let these people do their damage. Regent, here is the map—it's all we have, and I know you already have a copy as it was an archived newspaper from the Phoenix Gazette in 1909. This map was created by the Smithsonian-funded exploration of this very cave. I am surprised you hadn't already removed all the relics before now." Jeff shoved the rolled-up map into the Regent's hand and slung his pack over one shoulder. "I will be meeting with the board when I return."

The uniformed officers practically shoved them out of the cave entrance and onto the shuttlecraft. Before they even had time to discuss anything among themselves, they were on top of the mesa at the camp. There was evidence that the data van had been rifled through as if they hadn't believed they were forthcoming with all the data—which Jeff knew he personally had not. But, what of his find—the journal? Had they discovered its whereabouts?

# CHAPTER 6

## The Force of Human Nature

There was absolute confusion on the faces of his students when they returned to the mesa. Jeff told them to forget the Regents and proceed as if nothing had happened. He promised to deal with the fallout. Dutifully, his students began recording all their impressions in their field logs. This was part of the graded portion of midterms, although it represented extra credit. These graduate students were set to receive their degrees in May.

Jeff recorded his impressions on his data card, preparing to make a formal record for the Regents. For some unknown reason, the exploration was officially over. As he started to retrieve the journal, he heard the sound of another transport approaching the camp. The Regents were en-route. He slid the journal back under his pillow and stepped out to meet the transport at the canyon's edge.

The transport landed, and three Regents disembarked. One of them bore the emblem of a chairman. This was an official visit, not just a drop-in to hasten their departure. Jeff became instantly guarded. He wondered, what could the chairman want? What would he request or demand? He wondered if the Environmental Protection Agency had somehow gotten their wish.

Jeff straightened his dusty pants and shirt, attempting to look as if he were now wearing formal attire. He extended his hand to shake the chairman's hand. Alas, the chairman did not reciprocate, leaving Jeff to feel suddenly awkward. He shoved his hands in his pockets.

"Professor Ryan," began the chairman, "we have come to inform you that the Board of Regents has decided to cancel any further exploration of this cave dwelling." He handed Jeff a document and stepped backward as though he expected some sort of confrontation.

Jeff opened the document and examined the first paragraph. It read:

The findings of the research department have determined that the site has been compromised. The petroglyph records, in combination with the hieroglyphics, indicate that someone with advanced knowledge of these extremely different cultural writings has taken an effort to tag important historical records with their names, Skylar and Noah. The evidence of this graffiti has negated any evidence you may find proving your Egyptian connection. We request the names of all your students who have had access to these historical writings. It may be that one of your students chose to dishonor your find. The Regents will hold a hearing to determine the punishment for the individual or individuals who have destroyed a historical monument. You, Professor Ryan, will also be called into question on your findings.

"I will need time to examine all the recordings, as well as translate every wall within the cave-dwelling," Jeff responded. "I am confused by your findings. None of my students have those names. I will come to the Smithsonian at the beginning of next week to meet with the Board. We will pack up and return to the University sometime tomorrow."

One of the Regents began to walk over to the van as if he were going to drive it out of the camp. Jeff rushed over and stopped him from climbing into the driver's seat.

"Hold up there. This van is the property of the University. It is my responsibility, and it will be mine until I return it, equipment and all, to the archaeology department. At this point, all documents, all original documents, are still the exclusive property of the University. You have all the recordings already," Jeff declared.

Jeff had a sudden realization that one of his students must have recorded something that proved a rip in time. Otherwise, the Regents would not have reacted this way. He had looked at almost all the recordings and did not see anything that resembled graffiti, nothing with those names. The only place he had seen the name Skylar was inside the jacket cover of the paperback book. It had been written in ancient Arabic, not pictograph or hieroglyphic, but the name did appear out of place, out of time. No one, other than himself, he thought, had seen the contents of the wooden toy chest. He wondered what exactly they had discovered. It was a mystery.

The Regents reluctantly returned to the transport and left the area, stirring up an enormous dust cloud that covered the entire mesa. It was a white-out for minutes. Finally, when the transport was no longer in sight, the students rushed towards the Professor, each with questions, all talking at the same time. It was temporary chaos.

"What the hell? None of us would have marked up a historical find. We are professionals, for god's sake," Sid yelled over the other students, who were probably all expressing the same sentiment.

"Look, I know that none of you would ever do anything to harm an archaeological site. I don't think anyone did, now or in the past. It is my impression that they found what they wanted to know and didn't want to continue to fund us. Nothing more than that. It looks like this adventure is over, for now at least, but I, for one, am not done examining this find. It is early enough. If any of you are willing, I would like to make one more attempt at recording more," Jeff proposed.

Sid and Sally jumped at the challenge. Frank decided to stay with June, who was having issues with staying even one more night.

After the dust cleared from the canyon and some of the mummies had been airlifted out of the cavern, it appeared they were alone. That evening, Jeff, Sid, and Sally returned to the cavern. They recorded everything they could that hadn't been in the original plan. Each new find was documented and mapped out for the next time. It was after dark when they finished and decided to stay the night. Jeff radioed up to the mesa, telling Frank to hold down the fort. They would be there in the morning.

Jeff positioned himself again in front of the doorway which held the toy box. His mind was a whirlwind of confusion about how he could conceal the box, the time anomaly, from the Regents. After Sally and Sid fell asleep, Jeff moved the box up to the secret floor above the room. He dusted the area to conceal any footprints left leading to or coming from the stone wall stepping stones. This helped to conceal not only any movement of "relics" but also showed that the wall was just a wall, not a pathway upward. He fell asleep and dreamed of the messages he read inside the pyramid.

His sound sleep was again interrupted by the sound of heavy equipment and what sounded like boots hitting the stone walkway. He woke with a start.

"Oh God," he thought as he jumped to his feet. He held his position, frozen. "The Regents. But why now? They just left. Why would they come back now? Had his students told them they were inside the cavern again?" His mind was still partially numb from sleep.

Sid and Sally woke with a start as well. Rushing down the main corridor was what appeared to be officers. Jeff rubbed his eyes to make sure he wasn't dreaming. They approached three by three down the main hallway from the stone-cut cave entrance. In the background, there was the sound of a large hovercraft. The troops surrounded the still-prone students as Jeff rushed to protect them. It all seemed abnormally violent, odd, not natural, almost warlike—a kind of violence he had read about in the journal but not in his world, his memory.

"What is this about?" Jeff demanded from the uniformed officers. As he inspected them, he noticed they wore an emblem that read, "Relic Retrieval and Restoration Unit." He had never heard of this organization. If you retrieve a relic, why would you restore it? Antiquities don't need restoration. His mind was flashing a hundred thoughts all at the same time. Before anyone could answer him, the troops parted, and the Smithsonian Chairman appeared.

"Professor Ryan, we are here to remove the statues within the corridors. We will also be removing part of the golden wall. These men are to be given your full cooperation. Do you have a map of the area we need to inspect?"

"Kids," Jeff said, turning to the students, "the Regents are coming here a bit prematurely to remove the gold statue we recorded. I informed them that we hadn't had time to explore, nor had we found anything interesting yet." Turning to the Regent and then back to his students, Jeff shrugged his shoulders. "Look, I get it that you think we have violated this dig. You already have what little information we have in the way of a so-called map. My graduate students still need time to prove they will be of value to the Smithsonian as archeologists. I still need time to prove that we haven't spoiled this archeological site and we all could use the time to sort out any problems. Could you and the board be persuaded to extend your removal of the relic until at least another two weeks to give us more time to record?"

"NO, the Regents require the gold. I mean, the Smithsonian requires the statuary to prove to the Environmental Protection Agency that this exploration was of historical value. We will be removing the Asian statue. It has been decided that these men will dismantle it—with care—and remove it this very day. You and your group will have to vacate the area. Is that understood?"

"Don't you think that any damage to the cave or the South Rim would be viewed as defacing a world monument by the Environmental Protection Agency?"

"No, we are professional as we assumed you were. You and all your students will leave now, today and not a moment longer. The Board will be anticipating your report. Good day, Sir."

Jeff backed down and conceded. The crystal city had to be left as it required too much time, and that was something they didn't have. Returning was risky and would not be sanctioned. Jeff worried that he might be barred from the University by the Board of Regents. What had his students recorded that caused this sudden halt to the exploration? The shuttle flew them to the top of the mesa again. The students were still sleeping. It was obvious they hadn't even awoken to the sound of the transport.

Jeff decided to come up with the equipment and the funding to continue this at the end of the semester. After the students assembled, Jeff cleared his voice. "So, any of you want to spend your summer up here, providing we can get access again, you are welcome. For now, let's pack up and get out of here before they return and force us to leave all the equipment."

The camp was struck in record time, and all the students boarded the van. The professor gave Sid the keys, and they were off down the dirt road toward the main highway. Clouds began to form over the canyon—from whip cream white to black as night. Thunder rolled over the mesa and echoed through the canyon below. Suddenly a downburst—rain, hail, lightning filled the sky. The roadway became slick as ice, the tires spun, and the van went into a spin.

June's scream was almost as deafening as the thunder and flashes of lightning as the van whirled wildly into a ditch. The heavy equipment flew through the air, landing on top of June's body. When the motion stopped—with the van lying on its side—Jeff crawled over to his youngest student. Blood rolled over the seatbelt, which held her in an abstract position. He reached her, shoving a large metal case off her chest and held her wrist, feeling for a pulse—nothing. He pressed his data card against her now blood-soaked neck and fell backward against the side of the van. Tears rolled down his face. She was gone—maybe it had just been her destiny to die. If Sid hadn't caught her in his arms as she fell in the cavern, she would have died in Cibola. The sounds of the wind whistled through the broken windows, crashing thunder, and hail striking the metal sides of the van were broken by soft moans blending with the fury of nature. Tragedy had struck; the adventure of a lifetime for one of his students was the last adventure she would ever experience in this lifetime.

As the students became aware of June's death, they each inspected their own bodies, checking for signs of trauma. Jeff located his pack and secured the documents and videos. He placed the drives in his pack and called the emergency response team of the World Park System. Within minutes, an ambulance transport arrived to take the passengers to the nearest clinic for inspection and care. The Regents were as prompt as the emergency team. They airlifted the crippled van up to a large transport and out of sight. Jeff was glad he had the presence of mind to remove the drives and documents prior to their arrival. It was odd, he thought, how quickly they had returned to confiscate the van. It was almost as if they had orchestrated the accident, but that was foolish, they hadn't created the weather that initially caused the accident—or had they? The technology existed but was deemed unlawful.

As Jeff looked out the window to the canyon below, his mind was filled with grief over the death of his student, June. She was so fragile, he probably should have left her name off the team from the get-go. She had reminded him of his daughter—brilliant, pretty, and insistent. Both of them had the uncanny ability to make him bend to their will; otherwise, June would have just heard about the adventure back in class. Her parents would be devastated by the news. To lose a child is one of the hardest things to overcome for any parent. His heart was breaking, just empathically feeling their grief. His heart was still attempting to heal after the loss of his beloved wife. It had been three years, but it still felt like yesterday. A single tear rolled down his rugged cheek, and he felt Sally's hand touch his shoulder. He put his hand over the top of hers—he didn't mind the kindness—it had been a welcome sensation.

Days flew by after the tragic accident. June's parents had planned a celebration of life ceremony to be held at the University's meditation center. All the facility, students, as well as several Regents, arrived to pay their respects. Jeff sent his regrets—he couldn't bring himself to attend. He knew June's parents would understand, and he had hoped his students would as well. The administrator had called Jeff regarding his lack of attendance and had given him his understanding. It was unusual for the attending professor not to appear at a service for one of their students. However, under the circumstances, it was excused.

Jeff had used his time inspecting the files of the dig, attempting to understand what the Regents had found that led them to cancel the rest of the expedition. The excessive hours he spent were both therapeutic and informative. He couldn't help wondering what the motivation was. He hadn't found any evidence of the so-called graffiti. Had they suspected him of finding the toy chest? Had they already had knowledge of this thing? If they had just allowed the further exploration of Cibola to continue, June would still be alive. Deep in his heart, he believed they had caused her death. He tried to steer his thoughts towards logic, not conspiracy, but to no avail.

The day before he was to appear at the Smithsonian for his hearing, he found a single video made by June in the east corridor number three. On one very small frame of the video, near the floor, were petroglyphs of a tall girl and a small boy. Beneath the pictograph were hieroglyphics with the names Skylar and Noah. Remarkably, there was also a very tiny cuneiform script of the same two names.

It was like a child's *smart board* where they were practicing writing their names. Was this the reason June had almost fallen to her death in crystal city? Was it the reason she died in the accident in the van? The tiny area of concern could have been easily overlooked. Had Jeff not been analyzing every inch of the video, he would have missed it himself. It was now all too clear that the Regents had found an anomaly and had stopped the information from getting out, at all cost.

What would his penance be for the knowledge? Terror filled his heart. He needed to hide everything, including the journal, but where? The face of his grandmother came to him. She was living in Sedona, a short trip by transport. No one knew her at the University; it would be a very safe place to go, to stay for a while. He called her, and she told him to come up. She would have a room ready for him.

It was almost eight p.m. when Jeff arrived at his grandmother's home. She opened the door, and Jeff was struck by her ageless beauty. She wore her hair long, and it draped across her slender shoulders. Her blue eyes were glazed with an aching look of love. It had been too long since he had visited. She opened her arms, and Jeff gave her a gentle hug. She looked graceful, even for a woman of a hundred years who still lived alone. Jeff's grandfather had been missing, presumed dead after a space flight to an unknown planet. She had survived without him for more than thirty years. Jeff took his pack, which contained the contents of the toy chest as well as all the documents and video he had removed from the van, and hid it under his bed. As he did that, he couldn't help laughing at himself. It was something he used to do when he was living in this very house as a young boy.

"Jeffery, what are you working on these days, and how is your teaching position at ASU going?" his grandmother asked in a motherly fashion.

"Well, it was going really well. I took my best students for an exploration of the Grand Canyon and ran aground with the Smithsonian Board of Regents," Jeff reported as he ate the last piece of biscuit smothered in homemade fig preserves.

"The Canyon? What were you doing there, and why were the record keepers upset with you?"

"Record keepers? Who are they?"

"Just a silly expression your Grandfather and I gave the Smithsonian when they took over all the history of the planet as if they knew the history better than the rest of us."

"I wasn't aware they had full control over all knowledge. That is what you are hinting at, right?"

"Well, right before your grandfather left the planet, the Smithsonian gained power from the World Alliance and began monitoring all the news, all the past history, as well as what would be released as news in the future. We believed it was a subversive power grab. If you control knowledge, you control the world."

"Funny concept, Gram. You sound like the book I'm reading. Speaking of reading, if you don't mind, I have a great deal of reports to go through for my meeting on Monday with the Board of Regents. I would like to just go to my room and talk more about this subject in the morning. You may not be tired, but I am feeling older than you this evening."

"Age is just a number. Youth is all in the mind and soul. Don't read too long, and get a good night's sleep. I'll see you in the morning." His grandmother cleared the dinner dishes and went to the patio to watch the setting sun. She had never tired of watching the red rocks change colors, matching the sunset across the sky.

Jeff returned to his room and retrieved the leather-bound journal. It was time to read, research, and discover the mysteries held within. He was struck with the thought that the journal writer sounded like the same generation as his Grandmother. Maybe this past history of violence was part of Grandma's early memories. He would have to ask her about that in the morning. He put on his white cotton gloves, opened the journal, and began to read.

# CHAPTER 7

## Milk of Human Kindness

Months flew by after my encounter with that kid. I had questioned my parents, but they seemed to just brush me off. It was maddening. They acted as if I had fabricated the whole event and that I was so inventive and imaginative that I should be helping to write their book. But no real information was forthcoming. I was in Junior High School, after all, and they were treating me like a wild child.

The holidays were coming. Halloween was just around the corner. I had dressed up every year and was planning my costume. I decided to be one of Buck Rogers's girls from the comic book and asked my mother to help make this happen. It wasn't until this moment that the odd teeter-totter of life came full force. I was informed that I was too old to trick or treat, too old to get in costume. The only way I could dress up was if I took my baby brother with me to trick or treat.

What good is it to look really grown up and sexy as a space chick if you have to drag your brother along? But concessions were made, and my brother agreed to be an alien for me. He had green skin and an aluminum foil warrior outfit on, antennas and all. I got to be not as sexy as I wanted as a Space Chick, but hey, *trick or treat*, right?

I marched my brother up the block on one side and came to the last house before the mountain started. As we began to cross the street, Seth appeared next to us. He was wearing a tight-fitting silver/blue uniform of some kind, and his eyes seemed even larger than I had ever seen. My brother stepped back and just said hello. Talk about being nonchalant.

Seth said hello and then reached for my hand. I just slid mine into his, and we walked to the next house. My brother went to the door and yelled, 'Trick or Treat,' as Seth told me he would be in all my classes after the Christmas Break. I thought that would be wonderful. It would give me the time to know all about this guy from "next door." When my brother returned, Seth looked at him with a strange expression and said, almost under his breath, "Little green men, talk about being stereotyped."

My father had been my confidant about this oddly handsome exchange student from Tau Ceti. Even though my parents, mostly Mom, had treated me like I had invented him, Dad wanted me to be sure to only show him the best of humanity. I had wondered why he had made that a particularly important point. I figured that was good advice for both human and alien visitors alike anyway.

Although I was always the one that believed everyone deserved the opportunity to experience the good with the bad, sometimes bad is good. How could anyone really appreciate the greatness in everything if they never saw the other side? I really believed people would take all things wonderful for granted without knowing the bad, sad, or tragic parts of life. This is, of course, coming from the all-knowing Junior High school student who hadn't really experienced anything terribly tragic in my whole life. So take that opinion for what it's worth.

As a blossoming twelve-year-old, it also seemed important to my Dad to remind me that although I thought of myself as all grown up, I was still his little girl. I should remember that he was always listening to my inner thoughts. Talk about being on your best behavior! I had always known that both of my parents were telepathic. I, on the other hand, did not think I had anything but the radio connected to my mind. I could receive it but not send it. What a bummer.

Jeff set the book down and laughed to himself. This girl reminded him so much of his daughter Athena at that age. It must be a timeless worry for all fathers of young girls. How to control raging hormones and still allow for self-expression? Jeff's mind floated back to the past as he and his wife Mindy raised their daughter outside of San Francisco. Athena was a beauty, as lovely as his grandmother must have been as a child. His beloved wife, Mindy, was struck down when Athena was just four. As a chemist, Mindy was working on a project that blew up in the lab, taking her life and the life of her young intern. A tear rolled down his cheek. As he regained his emotions, he thought to himself, "Thank goodness

Grandma was here for us, or I wouldn't have been able to continue on." He strengthened up and began to read again.

The New Year was here, and we honked the horn at midnight and went to bed. Not too much fanfare at our house, but oh well. With my absolute love of the holidays, the decorations, and the lights, it was equally a dreaded day when we spent endless hours taking them all down. It seemed to me like removing the stars from the sky. Soon the Christmas break would be over, and we would again go back to school. This New Year would signal the last half of the school year. Although I did love the time off from school, as did my parents, I was looking forward to seeing my best friend again. He had mysteriously vanished right after the fall break, and I had missed him more than I would have thought.

Seth had returned to school just as he had promised. Although I found him to be the most handsome and oddly beautiful creature to walk the earth, others only noticed his elongated head. They called him freak, alien, and weirdo. Because he had a gentle nature and a 'do no harm' philosophy, he was continually bullied for being "different." He never seemed to let it affect him. He just remained calm, almost too calm for my disposition. It would make me so angry to watch other kids be so cruel.

He was not the only person in my class who had been regularly subjected to this verbal torture. There was another girl from a very poor family who suffered from a skin disorder. They called her crater face and freak as well.

"Don't touch her, she has cooties!" the kids in class would yell. It broke my heart, and I would always go up to her to see if she was okay.

I would often find her in tears sitting on the floor in the bathroom. She had impetigo or something that I am sure medicine would have cured. It was one of the bad sides of Earth that we would not heal the sick unless they had the cash to make it happen. What about that Hippocratic Oath? I think the medical profession couldn't figure out the difference either.

Jeff paused, reading this section of the journal. Intolerance was something unknown in his world. His world always understood the dignity and worth of each individual as well as the gift to seek spirituality of one's own choosing. The notion that any child would be denied simple health care was unknown in his world, as was the idea that society valued material possessions over community responsibility. It was just not heard of now or in the past. Just odd. There was something about this journal that was screaming "time anomaly," "parallel universes." But how could that happen here on this Earth?

Here I was, the only girl in a neighborhood of boys, chased, captured, and scalped, never seeing the prospect of being "different." I was the girl from a family of telepaths, that weird family who could read minds. My gang of seven cowboys never seemed to believe I could read their minds because I couldn't. But that didn't stop their parents from spreading the rumors. I often thought of the Salem Witch Hunts and wondered when we would be burned at the stake. Even with all of that, I still thought of myself as just a normal human kid. No one ever bullied me. Or had they?

January 1963. With the tension in the nation and the stress of the students at school, cruelty, prejudice, and bullying seemed to be happening more regularly. Our small world within the classroom changed for a short time. I had thought that with all the tension in the country, we kids would learn to get along better, as we finally did in our room. The diversity in our small community at school included children with disabilities, language issues, culture, color, and religion. But we managed to form a bond that lasted for years. It was a time when people like George Wallace called for segregation today, tomorrow, and forever,[1] breeding hate and intolerance between the races. Our generation could and should become the voice of reason, the voice of peace, and the instrument for the world to love one another. Taking small steps, one person to one person, one classroom to one classroom, one block to a city, and to the world. But, oh well.

The turbulent year 1963 brought a lot of interesting spiritual, moral, and mental challenges into my life. It was during this year that I discovered my inner strength, my

---

[1] George Wallace used these statements in his Inaugural address while taking office as governor of Alabama

psychic abilities, and my sense of justice. With the world seemingly on the brink of war, the school carried out the weekly "duck and cover drills" in anticipation of missiles flying overhead. This practice began with the Cuban Missile Crisis and was still the news of the day, with President John F. Kennedy "standing our ground," telling Cuba to remove the missiles pointing at our shores. Fear and suspicion filled the air. The politics of the students seemed to reflect the opinions of their parents. Although October 1962 had passed, it was still only a few months ago.

My grandfather had been invited to the Manhattan Project[2] nuclear bomb test site in 1945 but chose not to go due to religious reasons. My father remembered seeing the mushroom cloud in the distance. It was a strange and frightening time to live in, for sure.

During this year, Seth finally began to reveal the truth about himself to me. He was sent to learn about the human condition. Seth explained that because humans were advancing in technology and had sent satellites into the atmosphere, the council needed to know if we had advanced in enlightenment as well. One of the planets within our own galaxy had been subjected to self-annihilation because they were unprepared to control or truly understand the technology they obtained. This planet housed the Science

---

[2] Manhattan Project, July 16, 1945, out of Los Alamos, New Mexico exploded Trinity an Atomic Bomb with the explosive power of 18,000 tons of TNT

Academy for the Council of Worlds. A small group of radical scientists began a war that eventually blew up their planet and stripped the atmosphere from Mars and even the moon.

Seth remarked, "Just to let you know, millions of years ago, what you call your moon was really an artificial biosphere, a sort of orbital planet. You were what the council called a binary planet. Earth and the moon had life, civilizations, and advanced technology. All this had been wiped clean after their planet exploded. The council needs to know whether humans in this time were more careful, more peaceful."

"What do you mean in this time?" I studied his face, but in the back of my mind, I remembered this story from my parents, that book they were writing, and wondered.

"Life here on this planet has been through many time eras. The watchers have been logging information on the advancement of people here for eons. I am here as an observer, and it's really important that I understand the emotions of the planet. My mission is to see if the children are advancing, becoming more enlightened, our generation to be exact," said Seth in an almost melancholy tone.

Startled, Jeff sat up. "That goes alone with the information inside the pyramid, the many eras. But what era were the hieroglyphs speaking of and why hadn't that information been part of our known history?" Jeff thought, "Maybe Grandma was correct about the Smithsonian controlling history, but why?"

Being subjected to bullying had been an important part of his education. It was to be used as a case in point for the council to intervene in Earth's advancement for the safety of the universe. The act of bullying proved the cruel nature of humanity. Seth made the case to the council that the fact I had stood up for the bullied people spoke to the advancement of some. The fact that I hadn't noticed that I was indeed different also helped the cause for the growth of the human spirit. I was worried about my generation because I knew we had a terrible tendency to be cruel by nature.

He was on a mission to find out if humanity had progressed to the next level of enlightenment. His people and the Galactic Federation for the protection of the "Planet of Children" were our ancestors, our creators if you will. They needed to know if we were ready to join the main council.

The findings were not really in favor of this planet. They surmised that the violent genome was still controlling the majority of the world. The Cuban Missile Crisis did little to help our cause.

The almost catastrophic destruction of America happened when two hydrogen bombs were accidentally dropped over Goldberg, North Carolina. The military assumed that the safety measures put in place were the reason they did not explode. Seth said the council neutralized them remotely. They understood this to be a foolish accident of children and not an act of violence or war. I didn't know about this almost disaster. It was never on the news, and I still shake my head in disbelief.

The council has been watching Earth for billions of years. Earth, referred to as the "Planet of Children" by the Galactic Federation and the Council of Worlds, is seen as very un-evolved. So, when we think we are so smart, they, in their description of the inhabitants of Earth, let us know that we are not as smart as we think we are. In their eyes, we really are all children, no matter how chronologically old we are. They view themselves as our parents, our guardians. I'm not sure how that makes me feel.

Seth told me why his people thought he should come now. The earth had progressed in technology and weaponry to a point of danger to themselves as well as to this galaxy and the universe. With the first nuclear weapon exploded and the first manned space flight by the Soviet Union having been accomplished, it seemed an important task to have on the ground observers.

Seth was not the only foreign exchange student. He had friends from other planets, other galaxies, also here to observe and report. Seth told me that all the nuclear weapons, hydrogen bombs, and atom bombs had come from reverse technology obtained through unfortunate crashes of the aerial observers.

"The Soviets captured and saved a crash victim, one of the species of the observers from the Council of

Worlds during your World War Two. This small but very dangerous species, I think you call them the 'Grays', shared the technology of upper atmospheric flight and weaponry. All missile technology was obtained well before its time," Seth explained. He emphasized that this technology was years away from what we, as a developing species, should have figured out for ourselves. Oddly, during this same time period, a Pleiadian ship landed outside Washington D.C. and had a meeting with the President and his council to arrange a treaty. When the United States refused to agree to an end to the war, they simply left. At that time, the Galactic Federation and the Council of Worlds were not apprised of the Zeta Reticule violation of non-interference with Earth.

Seth told me it seemed imperative that the council send a child to see if the children were more advanced than the adults on this planet. With documentation from the observation science lab, the geneticists concluded that some younger children had obtained higher mental and cellular advancements. The council was unsure if the elders of the "Planet of Children" had simply not advanced enough to play with dangerous weapons. I was reminded of my father's warning to be sure to show the best of humanity to Seth. Did he know what Seth had been sent here to do? I know that my family had embraced everything otherworldly but this information terrified me, alien invasion terrified.

I wondered if he meant that children, maybe whole families, had been abducted and experimented on. I flashed back to the missing hour and wondered if they had tested my genomes, altered or damaged them somehow. It was a scary thought. Although Seth had been sent to observe and learn, document, and report back his impressions, he felt that the youth had potential to change the course of history. I thought about his comment about the other kid, the exchange student sent to take his place. Was he or she a pre-teen from the hidden city, which I guessed was Los Alamos? Had he/she been abducted from his family and was living on the observation ship right now? There were, as always, more questions than answers.

The year ended with a familiar goodbye from Seth, "I will be seeing you in the future. Our destinies are linked. We have much to accomplish. We share a great task in the future. We can change the course of humanity. Somehow I feel that it has to be you and I together, exchange student and human, who will finally bring the children into the new age of tolerance, peace, and love." Seth walked up the trail towards his portal and was gone.

"I am still just a kid. How could I, even with the help of my alien friend, the foreign exchange student, ever be able to change the world? He was so nice, remarkably handsome and gentle. What could he do with or without my help?" I daydreamed about this thought. I considered the future of my world and the exchange of cultures and ideas Seth and I shared throughout the year and wondered, "Could it happen?"

Jeff put the gloves inside the journal and closed it. His mind was on fire, thinking about what these two teens could have done to alter time itself. "Had they, with that mysterious destiny, altered life on Earth to the point where it was no longer history? Had they eventually altered the past and eliminated violence?" Whatever they did, it must have worked because there had never been that kind of horror on Earth. People worked together, each using their unique skills and training to contribute to the prosperity, health, and creativity of all who resided on Earth.

He was stunned at this thought. He hid the journal under his bed and attempted to sleep. Thoughts, like a movie trailer, flooded his mind. He watched the scenes of these teens. Oddly, the girl resembled his daughter and Seth reminded him of his uncle—only young, his Uncle Noah. He did know someone with that name. "How random," he thought. So what had that meant?

'Skylar' was the name on the paperback book and 'Noah' was the name of his Uncle. *'Coincidence?'* he thought. This knowledge could complicate his meeting with the Regents. His mind would have to seal this information, this coincidence, firmly away. Finally, deep sleep overtook him and the mind went still.

# CHAPTER 8

## My World Is Empty Without You, Babe

The early morning sun poked through the curtains and struck Jeff's eyes. He woke with a start, looking around as if to remind himself of where he was. He had dreamed of the family inside the inner Earth catacombs and of the author of the journal. Sitting up, he inspected the walls. The Native American artwork his grandmother had adorned his room with was illuminated by the soft early morning light, and he suddenly felt calmed.

He smelled fresh coffee brewing and the aroma of freshly baked biscuits. His Gram was the best at making those baking soda morsels. He dressed and found his grandmother sitting on the back deck, watching the sunrise over the painted mountains that surrounded the Sedona area. He was suddenly struck with the notion that she seemed to need very little sleep, unlike people of lesser age. She was an enigma, an ageless beauty not common on Earth. Watching her sitting on the patio chair, the sun sparkled in her white hair and she looked like an angel. Jeff smiled to himself as he sat down at the table next to her. As he sat, he suddenly felt a sense of melancholy. Mindy would have loved this view, he thought. He missed her.

"Jeff, stop living in the past. The future is always a fresh start. Mindy is still with you, just on the other side of this reality," his grandmother said, patting Jeff on his hand and giving him a knowing look.

Jeff wondered how she could do that. It was as if she could read his mind. She always had that gift, and as he grew up in her home, he had learned to hide his emotions, his very thoughts from her mind. He would have to practice this skill when he met with the Regents.

"How about some quinoa and fresh strawberries for breakfast?" she suggested.

Jeff had been staring out at the beautiful sunrise as his mind wandered into the past, rethinking the information he had read in the journal the night before.

"That sounds wonderful," he responded. "Say, I was reading an interesting fiction the other day when I suddenly realized it had to do with your early years. Could I ask you some really odd questions?"

His grandmother, now in the kitchen preparing breakfast, yelled out the window, "Sure, what was the book about? I love to read."

"It's just a handwritten journal I found. It has some really strange information in it regarding a very violent past starting around the 1950s."

His grandmother fell silent for what seemed an eternity. As she came out to the patio with breakfast, she had an odd expression on her face. She set the bowls down on the table and walked to the edge of the deck, looking out over the horizon.

"So, you had a dig in the Grand Canyon. Where exactly in the canyon were you investigating? It seems as though you have discovered something you were not supposed to know while on your expedition. Is that why you're concerned about the meeting with the Regents?"

"We went to the cave that was found by the Smithsonian in 1909 that has been sealed since. Why do you ask?"

"Was that where you found your fiction novel?"

"It was. Why do you ask?" Jeff had that feeling again that she was reading his mind. How does she do that, anyway?

"Before you go on, I think we need to have a long conversation about life, history, and living in the now, not the then. I have a mind-blowing revelation to tell you, which you will have to keep secret, just as I have had to do for the past 86 years or so. It's a heavy burden which I feel bad about placing on you, but it seems it's now imperative you know the truth."

"I don't understand. I did find that novel inside the cave of the Grand Canyon and I wanted to find out what history you remember from around 1960, as the information in this novel has a lot of information that wasn't known scientifically at the time. It also had a violent history that we did not have. What is your connection?"

"Was it a leather-bound book?"

"Yes, I found it in a wooden chest with a child's book, Snow White and the Seven Dwarves. It had the name Skylar inside the cover but it was written in Arabic."

Tears formed in Grandmother's eyes and she slumped to the table. It seemed that Jeff had hit on a memory of something very sad. It broke his heart to see her cry.

"Skylar was my firstborn daughter. She died when she was barely 15. I think you should bring me the journal. How much have you read?"

Jeff was suddenly dumbstruck. What he had read in the journal, not to mention all he saw inside the cave, had his mind stretching to grasp reality. What is reality, really? He thought to himself. "I'm not too sure I should show it to you. I haven't read very much but it seems very dark. You have already cried just because I said Skylar, and I don't want to stress you. You aren't that young and I don't think your heart can take it, Grandma."

"There is too much for you to know. Maybe you should read on for a while and then we can talk. I will need to see the journal and to know what information you have gathered so far. I am going to lie down a while and I will see you at lunchtime."

Grandmother left and went to her room. Jeff grabbed another cup of coffee and went back to the journal. No need to hide it anymore, so he sat out in the yard under a tree and began reading again.

# CHAPTER 9

## High School Trek

*September 1963:* Seth did not return to school, and I spent the remainder of my time in Junior High without my best friend. I missed Seth all the time; he was my rock, my acceptance, my fearless friend.

One day, I was looking for some of my rocks and crystals to make a necklace. My hand glided over the collection, and suddenly there was a hot sensation on my fingertip. I moved the large quartz crystal aside, and there it was. It was the shiny silver rock I found when Seth had just appeared in front of me on the trail years before. I picked it up, and it seemed to begin to warm even more in my hand. I thought of Seth, and then I heard his voice in my head.

"Hey, it took you long enough. That is my gift to you to help us communicate over time and distance. Make this into a pendant and wear it for me. You will understand why later."

I mounted the stone, which my father said appeared to be a shard of a meteorite melted into a small piece of Mica. I wore it on a long silver chain. It became my lucky charm and my talisman against evil. The little stone would warm and cool all through the day. I had often wondered if it was sending impulses to Seth, letting him feel my emotions each minute of the day. That was the only way

I imagined this stone could communicate across galaxies. We were somehow linked. Had it fallen off his craft? Did he leave it for me to find? Or could it have been radioactive? After all, I did live downwind of the Los Alamos nuclear mission testing grounds.

The years leading up to High School saw history in the making, both good and bad. President John F. Kennedy was assassinated. Some moments in your life are super glued to your brain—this was one of those events I would never be able to put out of my mind. I was on my way to the office due to disruptive behavior in the classroom. Okay, I can hear you thinking, "so you turned into a trouble maker." Really, here is what happened.

I was without my best friend, Seth, who acted as my filter, keeping me from spilling over with emotions. Seth was also part of the Protection Force, and I was just a puffed-up little mouse pressed against the wall with a cat at my face. I looked tough but shook inside.

I still felt the need to defend the helpless and abused; it was my mission, you see. A boy called a classmate stupid, and I stood up and told him, "Knock it off, you creep!"

"Go to the office," the teacher yelled, and without a pass, I was out the door. I never arrived at the office. An announcement came over the PA system.

"Teachers, instruct your classes to gather their belongings; school is out for the remainder of the day.

President John F. Kennedy[3] was shot today and presumed dead." I could hear the school secretary's voice crackle as the announcement ended. Silence fell across the hall, the school, the country. I stood and cried, alone in an empty hallway. The pendant resting on my heart seemed to burn. Why had someone killed a loving family man and President? The world was too violent. It felt as though Seth would be the world's undoing—we were too untrustworthy as a species. What would they do? Destroy us? Enslave us? My mind spun in terror—a silent knowledge of the observers and the fate of the planet seemed to be hanging in the balance.

The civil rights marches and race riots were in full swing the summer before my first day in High School. My uncle, a minister in the Unitarian Church, went to Birmingham to march with other faith leaders supporting Martin Luther King. I was very proud of him. How do you tell your friends that your uncle had been arrested with Rev. Martin Luther King and a gang of ministers and still sound really proud? You don't.

The Beatles were the popular band, and Sonny and Cher were topping the charts. On May 14th, the second atom bomb[4] was exploded by the Chinese. The United

[3] President John F. Kennedy was shot and killed November 22, 1963 in Dallas, Texas by a lone gunman during a parade.

[4] October 16, 1964 China exploded their first Atomic Bomb declaring it would be their last...done to prove self-defense in an atomic age

States conducted another above-ground nuclear blast at the Nevada Test Site. Seth had gone home to report to the council that he felt humans had many great traits and that he still believed the children would bring the world into peace. I had wondered if he had sent the wrong message. Did he know what had happened to the world when he left? Was it his mission to help or to hurt the children of the earth? I worried every night as I stared out the window at the night sky, wondering what light in the sky Seth was occupying.

The beginning of High School in 1965 was a strange time for America. The youth of the country were going through a metamorphosis. Rules of conduct and rules for law and government were being questioned. The mini skirt made its fashion appearance, and I tried to get my mother to let me wear one to school on day one. Mom was a futurist thinker most of the time, but when it came to my appearance, both she and my father were in agreement that I should look like a refined "young lady" at all times. Give me a break.

I had purchased a really nice mini skirt at a J.C. Penny's bargain basement sale. I put on that skirt with tall white boots and a sweater seat, and I went into the kitchen. "No way will you leave this house in that outfit. Go and put on something proper," my mother stated in her teacher mode.

I changed my skirt and boots, and it was a good thing too—half the girls who did wear the newest miniskirts were sent home to change. There was a strict dress code in force. Although the young wanted change, it did not come easily, and some things were just set in stone—like the Stone Age.

My first day seemed really difficult. I had to race down the halls to make it to my classes. I got lost two times and had to ask another student, who really had no clue either, how to get to my next class. He was going to the same unknown location, so we simply wandered the halls together. It was a tough day, but afternoon came, and I went home.

"The Outer Limits" was on and the new "Star Trek" had just been announced. Although my parents still climbed the ladder to watch the skies, had their attention riveted by these new television shows. Little did we know that the television would, in time, become our keeper, enslaving us to sit patiently and watch life but not live it.

By the end of the week, I had finally gotten a handle on navigating and keeping my schedule. I ascended the main stairs to the second floor, rushing so quickly it felt like a dash to the end zone in a football game. As I rounded the corner towards my geometry class, I collided with an incredibly tall guy.

I hit the floor, my books scattering across the hallway. I felt embarrassed, angry, and pain surged through my body. I looked up at this towering basketball player, my gaze traveling from his body to his face, and a scream burst from my mouth.

"Seth, oh my God, it's you," I yelled. "It's really you!" My voice seemed to echo down the hall.

Seth extended his hand towards me, and I reached up to him. Before I knew it, I was standing. He was as beautiful as I had remembered, but he had gotten taller. I had to stretch my neck to look into his deep blue eyes.

"Anna, I told you we would meet again," Seth said as we began to gather my books and papers.

"You know, Seth, I am surprised you remember my name. I don't think I ever told you what it was, although I know you heard it during roll call at class," I whispered as I straightened my dress.

Seth smiled, and his bright blue eyes seemed to twinkle. "You are Anna Cronus, daughter to parents of Titan's Science Academy, now relocated. You are the smart, witty, pretty founder of the Protection Forces and my best friend." With that, he walked away.

Jeff dropped the book on the floor. "Oh my God, I was right. This is my family. The journal was written by my grandmother."

"Grandma!" he screamed, "Come here right now. I am in the backyard. We need to talk this instant."

Anna emerged from her room, a look of dread on her face, as if she were being called to the principal's office. She knew she had a lot of information to share but she wasn't sure how much to disclose, especially before he was to meet with the Regents. She needed to gauge how much he knew before revealing the alternate history of her life. She sat on the lawn furniture next to Jeff and simply asked, "So, how much do you know?"

"Well, I got to the part where your name came up. I think it was my grandfather, Seth, who said your name on the first day of high school. How much of what I have already read is history that didn't happen? How can this history be true if no one knows about it? How can this much violence exist in a world I have no knowledge of? Or was this just a gruesome fiction you wrote in your teens? And how did it get to the Grand Canyon?"

"That conversation will need to happen after you meet with the Regents. I think the less you know, the safer you will be. I would like you to give me the journal. Don't read another page until you've settled your dispute with the Smithsonian and the Regents. It will give you time to prepare for Monday. There is too much information in the journal that cannot be seen by anyone. This journal proves a time paradox that would rattle the cosmos if revealed. So, please do this for me."

Jeff wasn't happy with the answer, but he went in and wrapped the journal up, marking the page he had stopped at with his glove, and handed it to his grandmother, Anna. She was right. He did need to work on an explanation to save his job, his students' careers, and hopefully his ability to return to the Grand Canyon cave in the future. He now knew who Skylar was and that Noah was his uncle. The case was closed, at least for him. But how would he justify that so-called graffiti to the Regents?

"Gram, before you go to bed, I have a question for you. It will help me with the Regents tomorrow. There was a video made by one of my students, June, who died during the drive home. She had a frame that showed a tiny petroglyph with Skylar's name, and cuneiform letters with the same names. It was very small and near the floor. Can you explain this?"

"Oh my, that is unfortunate. It was Skylar teaching Noah how to write his name in different languages. There should have been Arabic as well. You may ask if they carbon dated the names because they would date to around 1903. Maybe that will help, or perhaps it might hurt your case."

"That is amazing information. 1903, you say? That is going to make for an interesting conversation. I will have to figure out how to present this so that we can continue our research in the cave dwelling at the Grand Canyon. I'm leaving tonight to get a little sleep before I meet with them. I think I'll blame it on the 1909 expedition of the cave. Thanks. I'll call you when I'm done with the meeting. After I've finished the investigation against myself and my students, I'll return because we have a lot more to discuss. I have to continue researching your diary. I love you, Gram. Thanks for always being here for me."

# CHAPTER 10

## Do A Little Side Step

Jeff drove back home to Phoenix, spending a restless night there before finding his inner strength. He missed the biscuits and jam his grandmother used to make, but he figured he couldn't really hold any food down. He was nervous, to say the least. Pulling together all the strength he had, he strived to be as crisp, on point, and persuasive as he could muster. His career, as well as his students' futures, relied on him.

After downing a third cup of coffee, he boarded the shuttle to the Federation Building in downtown Phoenix. Once there, he gathered his papers and mustered his strength before entering. The same young lady was sitting at the reception desk. Looking up, she simply waved him towards the elevator.

"You know where to go. Good luck."

"Wow," he thought. The news travels faster than the truth. Hopefully that didn't mean that everyone had a skewed opinion of the archaeological dig in the Grand Canyon. The marque didn't have anything on it that there was nothing news-related to him, so he started to feel a bit better.

Jeff entered the lift and went to the 12th floor. The main Regent, the original one who had Jeff sign the document with DNA, greeted him.

"Welcome, Professor. Please go into the conference room through this door." He indicated the door on the far left of the desk.

Jeff went in. There were a group of ten Regents sitting around the oval conference table. They pointed to a chair at the head of the table, against the far back wall. Jeff thought to himself, "It's as if they thought I would bolt out the door if I were at the other end, Jeff thought to himself."

He sat and opened his briefcase, bringing out copies of the report they had sent him. He straightened up in his chair, cleared his voice. One of the Regents lifted his hand as if to silence him. Jeff just looked at the Regent, maintaining a calm and controlled demeanor.

"We are here to discuss the willful desecration of a historical site. The graffiti on your students' video feed. What do you have to say in regard to this serious breach of protocol?"

"First, let me tell you that my advanced archaeology students maintain strict ethics in regard to any historical dig. That's one of the first things they learned in their first year. Second, I viewed every video the students created and did not see what you called time anomalies and did not see so you must have. I had the section you sent me expanded by hundred times to find that image. Have you thought to the carbon date of that petroglyph? I'd bet it dates back to the 1900s and is likely could be a tag by Professor Skylar Adam Noah Jordan, who went by S. A. Jordan. I view this as an excuse to halt the research within the cave, as was done after the

Smithsonian expedition in 1909, when G. E. Kincaid and S. A. Jordan first reported the findings. In this day and age, it is a mystery to me why you would not wish to uncover the world's spiritual and historical records stored within that find. It reveals a lot about early history. Or is that the problem? Is that knowledge what you fear?"

"We currently have a team within the cave-dwelling. I'll instruct them to carbon date that right now." He pointed to the record keeper, who typed the request to the team at the Grand Canyon. "With what you've sent in so far, have you found anything that indicates a time anomaly?"

"No, but it does prove that the Zuni and Hopi mythology regarding the Great Flood holds some value. It appears that, much like the Library at Alexandria, it served as a storage area to preserve ancient spiritual traditions from around the world. That would explain why there is an image of the god Ra in the same location as the Buddha, and a combination of cuneiform, hieroglyphics, and petroglyphs within the same dwelling. It would be a shame not to allow me to continue my research for the Smithsonian Institute. I understand that you are in need of gold for further expansion within the cosmos, but don't be so quick to melt down artifacts before they are cataloged."

A small ting sounded, and the record keeper looked down at the screen. He stated, "Professor Ryan was correct. The petroglyph we assumed was graffiti was indeed dated to the early 1900s, so the students were not responsible."

"Well, it seems that the student who created this video is off the hook. You'll be able to tell that student that there will be no derogatory report filed."

"No need to worry about that. She died in the car accident that occurred on the way off the mesa. It was caused by what I'd call a very freaky and abnormal storm. They just held the Celebration of Life for her this weekend."

The Regents all began mumbling amongst themselves, and Jeff tried to ascertain what they were discussing. "Is there anything else I need to clarify, or may I continue with my research? My advanced students are expecting to receive credit for the completion of the dig and the subsequent research papers."

The head Regent stood up, which somewhat alarmed Jeff. However, he remained seated as did the other Regents. A decision had been quickly discussed while they mumbled.

"You have a week to present your new proposal to the board. We'll meet back here next Monday at 8 A.M. We'll review it, and you'll have your answer by that Wednesday. Is that agreeable to you?"

"What was wrong with the original agreement?"

"We want to ensure that you limit the expedition of your students to the main floor of the dwelling. I think you understand what I'm talking about. Then we want you to send a confidential report on any anomalies you've found during your own private explorations. This report, of course, will never see the light of day, and I think you'll eventually understand why."

"Fine, I'll see you all next week."

Jeff gathered all his paperwork and left the office. As he landed on the main floor, the young lady at reception looked up at him, and he gave her a thumbs up. She smiled, and Jeff winked back at her. He felt vindicated, but he was left wondering, what exactly did the Regents want him to reveal? He had a week to figure it out. With the day still early, he decided to have lunch before driving back to Sedona.

It was afternoon by the time he arrived, and he had neglected to call his grandmother. He stopped at the Native American Museum to gather further information on Hopi and Zuni mythology, particularly regarding the Anasazi, Anunnaki, and the Great Flood. He was lucky enough to have run into one of the elders of the Hopi tribe and got an opportunity to ask a few questions. The elder told him a story he had heard: the Anasazi were here on this land before the flood. They had learned to farm in the dry desert and about how to create cliff dwellings. A few were spared during the Great Flood. The Anunnaki sent a ship across the ocean to the Middle East and guided the Zuni and Hopi tribes to this continent and into the caves of the Grand Canyon. They lived underground for generations before emerging to the great mesas where they built their homes. The Anunnaki taught them again how to plant and provide for the tribe, planting maize with the dry terrace method, not unlike the hanging gardens of Babylon. "Interesting," he thought, "not sure how this will come into play for my research, but it's something."

He arrived at his grandmother's home, and she was standing in the driveway waiting. Jeff thought to himself, "She must read my mind."

"Glad you made it back, I could hardly wait. How did it go with the Regents?" she asked.

Jeff just smiled, she probably already knew. "Well, I got the problem solved by blaming it on Professor Jordan. Kind of made up his first and middle name, but I think that information had been lost to time because I couldn't verify his name. It became Skylar Adam Jordan, aka S.A. Jordan. They carbon dated it to the 1900's. Thank you for that information. You'll have to tell me how that happened."

"Well, come on in and have some coffee. I made cookies for you because you need something sweet to eat." They ventured into the house and sat at the dining room table. Anna went into her room and came back with the leather-bound journal. "Before you begin reading again, I will catch you up on how they found Skylar's and Noah's names. When we were hiding from…well that's another story. We went back in time to 1901, and we landed inside the cave dwelling within the Grand Canyon. As you know, there is an extensive library of world spirituality as well as galactic knowledge within the crystal pyramid. You did find that, didn't you? Anyway, Skylar was learning everything but decided to teach Adam how to spell his name in a multitude of ancient texts. That's how that got there."

Jeff was mystified. All he could do was down about four cookies and guzzle down the coffee. He didn't even have any question because…what was he supposed to ask?

"Anyway, enough of that. I am glad they cleared your students, and I am sure they will allow you to return to the dig. They can't wait to hide more history, but I bet they need to know what it is first. They, as an organization, are still to this day, very shallow. A generation of bureaucrats have come and gone, and this new bunch only knows they are protecting the world against fear. They just don't remember what that fear is. I bet they won't want to actually know once the reality of it hits. I know I am rambling on about things you are unaware of so I will simply say…read on."

Jeff was happy to just leave the journal on the table for now and just sit out on the patio and wait for the sun to go down. There wasn't a sight quite as beautiful as the sunset across the red rocks of Sedona. It would be particularly stunning with the clouds that had appeared on the horizon. There was something very spiritual about the entire area of Sedona. It wasn't the vortexes that existed around the area but that it had been sacred ground of the Native Americans. He was finally totally relaxed.

It was just before sunset when his grandmother brought out dinner to the patio. Anna always sat there at this time of night. It was as if she gleaned extra energy from the sunset. She had made cheese enchiladas with rice and beans. Very southwest but also vegetarian, not vegan. Anna had been a vegetarian for decades and it must have been healthy because she was remarkably fit even being 100 years old and counting. She had added two glasses of beer to the meal. What could be better? What a great evening, Jeff thought, as he finished off his plate and stretched with a big yawn.

Anna looked at her grandson with love in her eyes. "Honey, you look like you are beat. Do you want to call it a night?"

"You know, I think I will go lie down, maybe read a while. We will talk tomorrow. Goodnight, Gram." Jeff planted a big kiss on her cheek and gave her a half-hug, then went to the dining room to retrieve the journal. He planned to read until sleep took him. He walked into his room, changed into his shorts and t-shirt, and got into bed. Yawning, he opened the journal where the glove was holding his page.

> Seth smiled and his bright blue eyes seemed to twinkle, "You are Anna Cronus, daughter to parents of Titan's Science Academy, now relocated. You are the smart, witty, pretty founder of the Protection Forces and my best friend." With that, he just walked away.

Jeff understood why he had that strange feeling that the journal was somehow connected to him. Gram's name was in this journal. He wanted to call his grandmother into his room and interrogate her but thought it better to just wait until morning. He went on reading the journal.

• • •

Why does he do that…drop strange comments and information on me and just leave? My parents are from where and are what? Am I human or am I alien? Do I need a green card now? Give me a break, who drops this kind of bomb and just walks away?

I have to tell you, the last class was the longest in history. I could hardly wait to question Seth or rush home and talk to my parents. It would, of course, answer a few million questions I had about my parents, the roof, the paranormal gifts, and my love of science. I know I am weird, a nerd, to say the least. No tape on the bridge of my glasses nerdy, but kind of weird, alien maybe.

Seth could make me so mad, but I had to admit, just like the song by Sonny and Cher, my world was empty without him. I was so glad the council sent him back to me even though I was terribly afraid that they would still rule us too ignorant for the technology we had at our disposal. What would it mean for life on Earth if they ruled against us? Would we suddenly be under their control, their rules, and we, at least in the democracy, could lose our right of choice. All these thoughts were scrambling my mind. I had a shared destiny with him; he was here to learn and report. Was he really my best friend or was he just "the spy who came in from the cold"?

The evening I sat down at the dinner table with my parents and confronted them about Titan and the Science

Academy was like something out of a sitcom. I asked the questions point-blank and all I got in return were jokes about writing their book, that science fiction thing they talked about once in a while. I was suspicious of their answers and of this so-called science fiction novel. I wondered if Seth was telling me the truth or had Dad told him about the novel at one of the Science Club meetings Seth attended once or twice several years back. I hated to be made a fool of regardless. Was Seth just pulling my leg or spouting excerpts from "The Overseers of the Titans"? With my ever-growing knowledge of alien life, the truth of my heritage was a mystery to me. My parents seemed to just drop the subject and so I did as well, but it never stopped my wondering.

Our world was in a state of war from Israel to Vietnam with factions on all sides pushing political agendas that had nothing to do with the will of the people. The so-called conflict in Vietnam took up the news of the day. Fear of war and the rumor of war became the conversation at school. All boys nineteen years old had to register for the draft and everyone talked about their older brothers thinking about going to Canada. The official anti-war campaign was on in full force.

Seth seemed to be more concerned about space than the potential for war here. I wondered why his total concentration was on the sky. In my mind, the fact that the world seemed to be rushing into war seemed unintelligent and counterproductive to humanity's advancement in technology and a shared scientific curiosity. America was on the verge of sending manned space probes to the moon. Did the military want to use this technology as a weapon, not for exploration of space? The manned space missions seemed to be masking ulterior motives to me.

Seth and I continued to see each other and have many debates and conversations about the mission of the "exchange students" and the work the youth of the world would have to undertake. Seth once made a comment to me, "The children of Earth were to save our planet from sanctions, from being remade if we could not become more peaceful". When I asked him what 'remade' meant he kind of sidestepped the answer but then reminded me of Noah. I daydreamed about Seth and I on the arc and finally dismissed that terrible thought.

I was mostly terrified at the thought of our world's outcome. There was a dread, a sense of doom, knowing where we seemed to be heading. This was the same feeling I had when the air raid siren went off for the duck and cover drills, fear that someday the world would come crashing to an end. On one note we were advancing in technology, our fascination with space and the need to adventure out there, then the rumor of wars around the world taking the news off to hate, prejudice, and political turmoil. It was really more than my mind could ever comprehend.

The first week of June 1966, Seth traveled back to his ship and then on to meet with the Council of Worlds. When Seth left he hugged me and said, "Anna, I will return. I promise to send the best information I have gathered to the council. I know that you are a little different from most human youth but I am using you as the test for hope in the defense of Earth."

The thought of their ruling plagued me throughout the summer. Seth was gone and I missed him so much. The hot summer seemed to drag on as I took my walks up the mountainside each afternoon looking for Seth to happen by as I walked the trail up the side of the old quarry. He never came. The first debut of Star Trek arrived on September 8th, 1966, and the whole family was stuck to the television set. I could only think of Seth and wonder how much of this new program really represented space.

Labor Day came and the rush to get ready for the new school year began. I headed for my first class which was geometry, not my favorite subject. I was a million miles away, my mind was literally in the outer atmosphere, searching for the slightest emotional feeling that Seth was on his way back to me. I could never really understand the superglue kind of attraction I had to him, it was sort of unnatural. A week went by and I wondered if Seth had gotten really bad news from the Council and just decided not to face me with the truth. I was really worried.

Wednesday, September 14th was the start of Rosh Hashanah. Although my family was not Jewish, we always recognized all holy days from the religions around the world as a part of our Unitarian Universalist practice. Rosh Hashanah marks the Day of Judgment and the anniversary of the completion of the creation of the world. It's also known as the Jewish New Year. I found myself pondering the significance of this day, alongside the mission that Seth was on to report the "sins" of the Earth to the Council. Perhaps the mythology of religion is truly intertwined with alien interventions and alterations of life on Earth.

That morning, Seth had returned to school. I found him at the door of my first class. I asked him what he did during the summer, and he responded with a smile.

"The fellow youth observers and I spent the summer tearing up the sky around Earth," he recounted. "We almost lost the kid from Ursa Minor when he landed at the bottom of a canyon in New Mexico. We had to go down and give him a boost up and out of the dirt. We were seen and reported by the local sheriff. The Council was furious- we almost lost our positions here on Earth. We had to do a lot of talking to convince them that we were simply taking samples of minerals when the shuttle touched down."

Seth then looked around and noticed a curious student watching him. He quickly added, "And then I read for another part in this really creepy horror film. What a lot of fun."

I chimed in, "Seth does Community Theater during the summer. Want an autograph?" The student shrugged his shoulders and went into class. We were finally alone in the hall.

With a quick hug, Seth pulled me near.

"Good news," he said. "The Council has stated that Earth has a grace period of one decade to grow up. I hope that will be long enough."

We had the next decade to prove that we were ready to advance as a planet. I thought ten years sounded like a long time, but maybe to them, it was just days. After all, the judgment was official on Rosh Hashanah in ten days. But a decade? Perhaps time is warped in space.

Meanwhile, Jeff was struggling to keep his eyes open, even though he wanted to continue to read. He wondered if the Council had somehow changed time and history. Was it the Council of Worlds responsible for the different ages of man as written in the mythology of the Hopi and Zuni tribes? Sleep was upon him, and he soon fell sound asleep.

# CHAPTER 11

## Human-be-in

Jeff woke well-rested and refreshed for the new day. He had read a bit during the evening but thought it might be time to read with Gram. She would be able to narrate some of the weirder information to him. The amount of violence was breathtaking, but the information about a Council of Worlds monitoring Earth was something he hadn't considered, as well as the idea that they had somehow orchestrated the end of humanity in our pre-history past, or at least his pre-history past.

He put on his sweatpants and long sleeve T-shirt, remembering that it was usually cool in the early morning. Slippers on, he headed out to the patio where his grandmother was watching the sunrise as she did every morning. He sometimes thought she was like a cell, gathering energy from the rising and falling of the sun. As weird as that sounded, it might be true.

Anna looked back and saw Jeff coming out to join her. "So, did you get enough rest? I see you have the journal with you. Are you going to just read out here all day?" she asked.

"No," Jeff replied, "I thought we could read together so that I could ask questions along the way. It might make the information easier to figure out. I had a thought while I was reading last night that the Council of Worlds caused the destruction of different ages of man and wondered if they are the reason my history doesn't match your journal."

"Well, that is a loaded question. Yes, the Federation and the Council of Worlds did have a hand in the destruction of humanity but not just humanity, the reptilian age as well," Anna explained.

"The first age of Earth was that of a lifeless, shapeless world. With the addition of the moon satellite, the Earth had rotation which caused the planet to begin creating life. This period of Earth history goes back millions of years. Meters caused this destruction. It was the age of fire and ice and almost ended all life on Earth but the oceans retained life.

"The second age of Earth came when the fish began walking on the earth and soon became reptiles and some dinosaurs. The Reptilians had a civilization after millions of years. After a time, they had begun wars against the dinosaurs and the few mammals that existed on Earth. The Federation and the Council created a polar shift which created massive earthquakes, volcanic action as well as tsunamis. After that died down, the climate dropped and we had the Ice Age. It is theorized that a group of Reptilians built cities under the ocean floor to survive. They may still be there. I don't know.

"The third age was the manipulation of the creatures on this planet. Humans replaced the race of Reptilians on Earth. They thrived and again damaged the biosphere as well as creating war, hate, and unacceptable behaviors. It was a time when the great builds took place. The Atlantians were very advanced and created many wonders but also weapons of mass destruction. The two strongest civilizations were the Lumarians and Atlantians, then the Aztec and Mayan, and finally the Sumerian. The world had a boom of technology in that era of man. It was meteorites and the passing of the second sun that created earthquakes, volcanic action, tsunamis that ended this era of man. The 4th era was roughly recorded in biblical texts and other religious texts around the earth. The ending of this was the Great Flood. I think you know that history as a lot of that information is held within the cave dwellings in the Grand Canyon. Then came the fifth age of man."

"So, it was in the fifth age of man that time and history became compromised?" Jeff asked.

"Yes, now we are up to date on Earth's past. Shall we have breakfast and then begin?" Anna proposed.

Together, they sat on the couch in the living room and began reading the journal.

> "Early 1967 things seemed to be coming to an alarming head. The race war was in full swing as well as the Vietnam War. Riots, peace marches, inconsistencies in human behavior. It was enough to make me dizzy. The space race was on as well. So here we were, punching, burning, hugging, loving, marching for peace, and flying off planet in fiery rockets. Just weird, psychotic humans. I thought maybe Earth, as a planet, needed therapy, not

sanctions, or even the destruction from the Council of Worlds.

The terrible accident that took three astronauts' lives was reported and then hushed up. Gus Grissom, Edward Higgins White, and Roger Chaffe went down on a quiet page of history when they died in a fire during a test launch. Seth said that it had been too soon and that the world wasn't ready to venture into space. I had wondered if the Council or Federation had something to do with the accident. Maybe they thought, as I had, that the mission was more to spy on our enemies from space than to venture into the future of space flight. The alien force terrified me and I had way too much knowledge from an insider point of view to simply ignore my inner thoughts. As soon as the tragedy was hushed up, another launch sent Orbiter 3 into space, but no incident was reported. I wondered if the Council just ignored it like children playing with toys. Then the Surveyor 3[5] probe landed on the moon."

Jeff stopped reading and looked at Anna. "So, is it possible that the Council really did create those accidents?"

"I always thought so, but your grandfather would neither confirm nor deny. Who could say? The old history is full of very terrifying facts."

[5] Surveyor 3 sent to explore the surface of the moon on April 17, 1967

"The Council will be very worried about this," Seth told me as he read the article in the paper. "Not only is my ship up there, but the main transport vessel as well, and the entrance to the observatory is on the far side of the moon. I don't think the Council can let this continue on." Seth continued to tell me that it would take them time to figure out what the next move would be. I was worried.

In January of 1967, the Pow-Wow gathering of the tribes for a "Human-be-in" was occurring. Although psychedelic drugs seemed to be dropped from the sky as reported by the San Francisco Oracle, the youth created a new nation of emerging souls.

Although I was unable to attend this gathering, I read about it. The pow-wow both celebrated and prophesied the epoch of liberation, love, peace, compassion, and the unity of mankind. It all seemed so attainable. If the youth of the world, or at least in America, gathered in the name of peace, Earth had a chance.

Seth told me that some of his friends, junior observers, were at the event, and he had wished we could have been there. I would have loved the adventure. It just wasn't in our destiny to be there. At the time, I wondered why he continued to be focused on our destiny and the timing of some unknown future prophecy. The Earth, on the surface, appeared to be making enlightened advancements, but with the good often, the bad roars louder.

Hate was still raging, as was the civil rights movement. Martin Luther King had denounced the war publicly, assisting the anti-war sentiment in 1967. With the civil rights and anti-war demonstrations happening all around the country, and love growing, a coalition of hate mongers soon arrived. I would read the papers every day and wonder what the universe thought of our intolerance. We were all of the Earth and, therefore, we were all earthlings. So why should we find conflict over the color of our skin? Life went on and Seth and I grew closer. Despite part of my heart loving this beautiful alien, the logical part of my mind worried about his motivation. I wondered if this caution was born out of people's suspicions, growing up in a world of fear, the nuclear bomb, and the communist threat.

In April, even in the desert, spring was in the air. Peace was floating among the race riots as 10,000 kids marched in San Francisco against the Vietnam War. San Francisco became part of my imagination, and I dreamed of the day we would go to some marvelous march or peace of some kind. We were falling in love and often I forgot about destiny, dreaming only of a lifetime with Seth.

Seth and I planned on going to the Spring Prom. I was excited and, for a while, I forgot that Seth wasn't human. I designed my formal and Mom made it for me. It was stunning. The fabric was deep purple velveteen with an empire waistline, pearls separated the chest from the long straight formal. The back was almost nude, as low as my mother would allow it to be. On Friday night, Seth presented me with a tea rose wrist corsage, and we were off to the dance. It was a magical night. I told all my friends that Seth and I were going steady. Seth asked me what that meant, and I told him it meant that we were a couple and that other girls and guys were not part of our relationship. He thought that seemed natural because we were destined to be together.

In May, the school year was coming to a close very soon. I worried about that because Seth seemed to disappear for months during the summer. I had forgotten that he was just on loan to Earth, it wasn't where he lived, where his parents lived, or even if he had parents at all. How could we be a couple, separated by galaxies?

It was a Friday night, and we were on a date, watching a drive-in movie under the stars. The movie was "They Came from Beyond Space." Seth thought the movie was hysterical and laughed through most of the dialogue. I thought it was interesting—after all, it was about aliens who came to conquer Earth from their dying world. It sort of sounded a little like the tale my parents told of Titan and the Great War, about inhabiting Earth and other habitable planets in the universe. Seth just thought it was more than silly, but the information about the moon and the fact that the aliens had a hidden base there seemed to peak his interest.

Intermission came, and Seth opened up the sack Mom had packed with chips and Royal Cola, then handed me a bag.

"NASA announced the crew for the Apollo 7 space mission. It's reported to be the first manned mission to the moon. I will have to see the Council about this problem," Seth said.

"How long will you be gone this time, Seth?" I asked, almost in tears when I asked. We had become so close, I couldn't bear to be apart.

"I will be back on Monday morning. You won't even notice I'm gone," Seth held me in his arms and kissed me. His kiss always sent tingles through every nerve in my body, leaving me dizzy and more in love. He made me want more, but I knew this was wrong, even though it felt right. We left the movie early, and he drove me home.

As we stood on the front step, Seth again pulled me toward him and kissed me. Then, with a step back, he vanished in front of my eyes. He had never just disappeared before; he always walked to his portal. I wondered how that happened and why he could just vanish in an instant. I went inside. Mom sat on the chair in the living room and smiled,

"Seth is traveling farther than just his ship. He's traveling to the Council chambers."

"How do you know that, Mom?" I was stunned, just a little angry, thinking she had been eavesdropping on Seth and me.

"So, my great grandmother and grandfather are aliens? They were scientists from Titan? Does that make you an alien too? Am I a half-alien?"

"Look, the word 'alien' is rather insulting, if you must know. No, I am human. My parents were from the Lyra constellation, which was earth-like with the same biosphere. When they had to leave due to the war and the destruction of their planet, they came to Earth, to the satellite called the moon. HUMAN."

"We received a mental memo about the threat of discovery by NASA. Everyone up there knows. I'm proud of you for knowing your place, honey," Mom stood, yawned, and walked to the bedroom. I stood frozen in thought for what seemed like hours. Had she known how much I wanted Seth? Had she read my inner desires? It was unnerving. I finally went to bed.

Monday morning, I arrived at school. Seth greeted me with a hug, looking both ways to make sure no teachers were watching. It was not allowed to show affection on school property, which was more prudish than for moral safety, if you asked me.

"The Council has created an ambassador role for several of the inner council to meet with three power governments individually and then together to stop the further exploration and expansion of weaponry within this solar system, as well as the galaxy. One Zeta will go to Russia, a Pleiadian will go to the United Kingdom, and an Arcturian will visit the United States. I was a little surprised they picked the Zeta as an ambassador. That species created a lot of trouble giving the Nazis advanced weaponry during WWII. They would love to see Earthlings destroy themselves and give them back their home planet, as they claimed it to be."

"Their planet? Are they human?" I asked.

"No, they are neither human nor reptilian, but they are in league with the reptilian planets. Look, I will explain after school. We should get in and make an appearance anyway. The bell has already rung. I will meet you on our mountain, okay?" Seth rushed up the stairs and to the west wing of the campus. I dashed up two flights of stairs to my geology class.

That afternoon, I came home and talked to my parents about the ambassadors. Mom seemed really concerned, and Dad simply said, "Everything will be just fine. It is as it was supposed to be."

Sometimes, Dad could come up with the most abstract comments. I never knew how to take him…a joke, seriously or just placating? I dressed in my slacks and shirt and went up the mountain. I found Seth sitting on a large rock on the inner trail, out of sight of the city below.

"Anna, come and sit down. I've repelled all the snakes before you got here but left a tiny lizard for you to play with," he laughed as he handed me a tiny gray lizard. The lizard moved from one hand to another, it walked about as though it were traveling a great distance. I was struck by a thought: here I was, toying with a lesser creature. What were we Earthlings to the aliens? Nothing more than lesser creatures, toys to be played with and discarded? With that thought, I set the lizard on a small flat rock. Seth smiled as I placed the tiny lizard next to me.

"The Council sent ambassadors to the three major countries to make a treaty with Earth. I wish I could have been in that meeting. Can you imagine the look on their respective faces as the ambassadors from three different galaxies convinced them to sign an Outer Space Treaty?"

"This binding contract will help Earth and prevent nuclear weaponry from being shot towards the planet by different governments. But mainly, it was for the protection of the secret bases the Council has on the dark side of the moon. Earth agreed that it would be forbidden to place any weapons of mass destruction in orbit around the Earth. Any nuclear device was barred from being placed in orbit or on the moon or any other celestial bodies. It also forbade any government from claiming the moon or any planet as their own, as outer space was to be considered a common heritage of mankind. It was a very diplomatic moment in the history of the Universe."

At that thought, I knew my first impulse was correct. The government was interested in advanced weaponry, not space exploration. Who could make this up? If the governments of Earth knew and made agreements with aliens, why were the petty political wars continuing? Didn't we have other, larger matters to be concerned with? This time, in American history, was an ever-changing and always interesting blend of advancements in humanity, as well as the same old hate that has caused war since before recorded time. We, the humans on planet Earth, were facing an insurmountable force that could decide our fate. Shouldn't we, along with the three major power nations, finally come to the realization that we had to work together as a family, as one united planet?

Jeff stopped and realized that this was a part of his known history. It was the start of the New World Order where Earth began acting together as a planet, not as individual ideologies and political combativeness. But what was the other history written about? None of that was reality…at least not his reality. "Gram," he questioned, "I understand that you had a different history early on, but how did this one historical record happen in both timelines?"

"The question should be whether all the good history is the same in both timelines," Gram replied. "You will need to keep reading."

The Council of Worlds and the Galactic Federation were still having the observers report all the findings concerning the attitudes and advancements of the children of the Earth. Seth, as a junior observer or exchange student, was more than just a spy for them. He also was attempting to be part of the solution, part of the salvation of Earth. He had always thought that we, as a couple, could change destiny for the better.

Jeff pondered that last statement. Maybe they had somehow altered history. Perhaps they were the timeline fracture the Regents spoke of. Why and how had this anomaly occurred? In which era had it happened? He was more intent on finding the truth. "So, were you and Grandpa the reason for the time anomaly?" he asked.

"No, it will all play out as you read on," Gram responded.

Seth and I had become very close. We attended most of the rallies around the college campuses. We were still the "Protective Force," but we were beginning to change the mission more towards the path of peace. The peace symbol was adopted as a logo for the peace movement. It had originated in Europe as an anti-nuclear symbol. It worked anyway.

In 1967, large-scale, anti-war protests had happened in New York, San Francisco, and Washington D.C. However, the voices of love, peace, and understanding fell on deaf ears. Seth seemed ready to leave but remained, I believe, to be with me. I wondered if I would have left if it hadn't been for my family. The world was fractured. The fate of our planet was hanging in the balance, not by the violent nature of Earth, the threat of war and a continuation of more war, but from the total annulation of the population by way of the Federation. They could just wipe the slate clean, start all over. After all, it had been done before, just ask Noah.

Seth made an unusual observation about the small group of people who seemed pressed to undo the momentum of peace. "I wasn't told this, but I have an awful feeling the Zeta's are behind this."

"Who are the Zeta's, Seth?" I asked. "You never really told me about them. You mentioned they were not our friends and something about the Reptilians?"

"Sorry, I was just thinking out loud," he said. "There is a small counterculture within the Galactic Federation, members from the Draconian galaxies, the Zeta Reticulan and others who are intent on ridding the cosmos of humanoids. I always wonder, when there is no logical reason for hate, if they are somehow involved. You know, they are living among the population of Earth and are shape-shifters, looking just like humanoids."

That statement seemed to confirm my fears. I knew Seth was from Tau Ceti and that the Tau Cetians were both members of the Council of Worlds and a Galactic Federation member, but I hadn't considered a mass metropolis of alien species occupying the vast expanse of the cosmos. It was sort of mind-blowing. I pondered the question of outside sources causing Earth's occupants to be destructive, but I did see that we, as a young and still-developing species, had many bad behaviors to overcome. I also pondered how malleable little children are. Were we being programmed for violence by an outside source?

The civil rights movement was still raging. NASA was still sending missiles into outer space. The U.S. Ranger hit the moon, which reportedly rang like a bell for several hours. It made me wonder if the moon were a man-made planet, an observatory for all of Earth and the galaxy. It would have a great bird's eye view on humanity, if you could look past the space junk. Seth had called the moon a biosphere type planet. Did "type" mean artificial?

A solar observatory was launched into the upper atmosphere, adding to the space debris already present there. Earth, the beautiful blue marble in the sky, probably now looked like a park after a rock concert, littered with trash and junk. Despite the Outer Space Treaty, we, the Earth children, couldn't resist littering the sky.

Jeff broke into laughter. How true, he thought. That couldn't have been a more accurate description, even though he wondered if Gram ever looked at the Earth from space. The concept struck Jeff. Why would war bring prosperity? It seemed to him that the thought of war was so foreign to the moral of the world he knew that it seemed random. Again, why?

The Protective Force had now changed to the Peace Movement. Seth and I, along with our small army of Peace and Love, put up fliers and manned petitions, requesting the government get out of Vietnam.

"We don't want war; we want peace" was our motto. In the final quarter of high school, we managed to gather enough supporters to buy a bus to drive to San Francisco after graduation.

In April of 1968, Martin Luther King was assassinated, as was Robert Kennedy after he had won the democratic primary. A small but powerful hate group refused to have anyone in power that wanted equality and unity in the country. It all seemed so out of control. I was worried about the fate of our planet and still concerned about the Federation and the Council's reaction to our unquenchable need for violence and hate. I often thought I should tell someone, some government official, about the observers from the Galactic Federation.

But then, I knew they would simply think I watched too many science fiction movies and put me in an asylum. Seth never verified my fear that the aliens hadn't secretly planted the irrational hate to clear the planet of us messy Earthlings, but he did hint about species who hated humanoids. The universe seemed as complex and messed up as the children of the Earth were.

We had already mourned the death and assassination of Rev. Martin Luther King. His death marked a turning point in the Peace Movement. Civil rights leaders, peace advocates, and students from colleges around the country were now ready and willing to march on Washington to state our mission: PEACE. We just needed a few more "hippies" to join the cause, to gather and participate in a great experiment to soften the hearts of the government with the voices of children filled with love and peace. Even with my worry about the alien threat, I was full of myself—on a mission of peace, like Wonder Woman—single-handedly saving the Earth. Well, not really, but I was filled with an unrealistic empowerment of self-worth.

It all sounded so possible, so attainable. But was this a mission by the children of Earth, or was it a strange plot by the Galactic Federation? The last time I approached my parents about this alien thing, my mother simply said, "Anna, don't be so dramatic!"

It was more of a sidestep to the question, not an answer. I needed to talk to my parents about this Peace Movement, the upcoming bus trip to San Francisco, and then on to Washington D.C. I had to carefully pick my words so that they would allow me to go. I debated the methods of approaching the trip every night, endlessly in my mind. I never came up with a valid reason to cite. Finally, on the day I figured to be the best, I simply asked my question more like a statement of concern.

"Mom, Dad, I know you are not ready for me to fly the coop, but I am going on this trip with my friends and the Peace Movement to San Francisco and Washington D.C. after graduation," I stated. I did not want them to have the time to stop me in mid-sentence, but I paused for impact. "I just need to know if you think I am doing this for the Earth, or am I working with Seth and the Galactic Federation?" I formed my question as matter-of-factly as I could muster.

My mother rolled her eyes as if to tell me I was being foolish, but her mind was telling me that she was concerned that I was being manipulated by Seth. My father looked contemplative and, finally, because I was unable to get a read on him, he began to talk—or maybe a lecture would be more accurate. As I watched his face, I became worried that he would simply shut me down. He had a great way of making me change my mind without him telling me no.

"You, my dear, are 100% our child. Seth was telling you the truth. We came from what is now the asteroid belt. It was once called Titan, and yes, we were scientists, cadets, and council members of the Science Academy. We are human. There was a time in Earth's history that this planet, Earth, and Edam (the moon) were part of the greater Galactic Federation. Your mother and I lived on Titan before the Great War. We actually met at the academy and fell in love there. We used the time fold to come here 25 years ago—in this time, the ONE reality, the Now if you will—but thousands of years ago in that time. Have I blown your mind yet?" My father smiled that really silly Cheshire cat smile.

Jeff stopped abruptly and stared at his grandmother. "So, my genealogy is alien. My great grandparents were aliens from Titan thousands of years ago? Is that what I am reading here? And are you telling me that you were alive during the last age of man when all the ancient dwellings and obelisks were here to provide a worldwide web of energy? That Earth was a member of the Council of Worlds and Edam was occupied by the Space Academy. And how old are you … really?"

"That's enough. Really, are you yelling at me or are you asking questions? Yes, aliens, but humanoid. We're talking timelines and string theory here. Yes, Edam was a science academy, and guess what, it still is. Furthermore, I was born on this planet in 1950, so no, I was not here for the last age of man. That was your grandparents. I am my chronological age, although I hope to be here for a while longer. Can we continue?"

Jeff grunted a comment under his breath, but reopened the book, and they began reading.

> My father took a few more moments as if he were waiting for me to ask a question, but as my mouth began to form a word, his finger shot up as if to say, "I'm not finished."
>
> "You and your brother were born on Earth in this current time frame, and yes, I think your mission is for the greater good of the Earth. Seth believes he is also here doing work that will benefit Earth. We are just not too sure what the real motivation of the Council of Worlds or the Galactic Federation really is. Time will tell. Your mother and I raised you to use your gifts and your intuition to your best advantage. You know we monitor you, you know we can

track you, so you may go, but know that we are watching and waiting for your return."

I think my jaw must have been on the ground, gravity had intensified. I can truly say that the information had just floored me—really. Wow, that was way more information than I really wanted. It gave me more questions than I had already. So were my parents really, really old? How does that work? What is the ONE reality anyway? Was Seth also a product of this time folding thing?

Was he some million or billion-year-old kid sent here to make me like him—a lot? Man, talk about confusion! I would just have to let "destiny" drive the bus and see where it really took humanity, and where it took me—my destiny, my future, my relationship with this really old or young alien boyfriend. We were on a noble quest to save humanity from itself. What a superhuman endeavor that would be—or was it human? I wondered.

I really couldn't ask my parents all the questions I had about this time warp thing. It was like trying to visualize infinity; it never ends. My mind seemed unable to expand out far enough to even ask rational questions.

As for my genealogy, I realized I would have to spend time figuring out what to ask and how to understand the answers. For the moment, my focus was elsewhere. I was focused on my great adventure with my friend, my really good friend. We began traveling alone, alongside a bunch of other kids. It was a journey to unsupervised freedom. Wow

# CHAPTER 12

## Give Peace a Chance

Graduation had come and gone, and the summer of 1969 was upon us. The day was getting closer, and I was growing nervous. Apart from my day-long adventures up the mountain to the caves and quarry, I had never been anywhere without my parents. Some maverick I had turned out to be. Major doubt and concern filled my mind.

Mom had offered to help me get things together for the trip. I would have to pack a variety of clothes for climates ranging from California across the country to Washington D.C. How does one get a half a year's wardrobe, makeup, hair products, and shoes all in one suitcase? That was all the group decided we could take and still all fit on the bus.

"Look honey, you are going to be a Hippy, right? You can leave all the nice dresses home. Take two pairs of jeans, I had some really funky bell bottoms I had just purchased, one sweater set, one pair of tennis shoes, a long skirt and four t-shirts. Oh, and don't forget underwear for five days. You will have to just wash them out from time to time. Let's see... how about one pair of sandals and a poncho for warmth. You don't need makeup, you are beautiful just the way you are. And honey, one bar of Ivory soap should do it."

Talk about making a radical cultural change. That one suitcase represented the fashion crash of the late 60s, early 70s—minimization to the max. From fashionista to flower child in one easy lesson. How did Mom know how to do that? Oh right, she was a time dancer. Maybe Mom was in San Francisco in the past. Future. Whatever. Who knew?

We agreed to meet the bus at the football field and say our farewells from there. Mom and Dad packed my suitcase in the car. My brother, Mr. Nonchalant, simply said, "See ya," and sat reading his book in the easy chair. I blew him a kiss and he air-brushed it away. Oh well, I thought. He will miss me. Or not. Maybe he thinks he will get to move into my room. I wouldn't be gone that long, I thought as I walked to the car. .

Seth was waiting by the bus as we drove up. A big peace sign had been painted on the side of the bus, and the gang of seven was ready to go. We were off to Haight-Ashbury in San Francisco to join the world of "Flower Power". A large group had started a campaign called "Let a Thousand Parks Bloom". Our main mission was to recruit more young people to join us in the big anti-war rally in Washington D.C. We traveled by bus, lived on the beach, bathed in fountains and lawn sprinklers, passed out flowers, planted daisies in vacant lots, and sang songs.

I was bullied by girls who looked like I had just before I boarded the Peace bus. Well-dressed, college-bound, fashion-forward girls who thought we were some kind of anti-government radicals, hippies, deadbeats. A few years ago, I would have thought the same thing, but it still hurt. I wondered why they couldn't see the glory of the mission, the urgent need to build peace, not war in the world.

Seth's hair grew long and began to make waves that flowed down his back. His white skin never tanned, but he now sported a bandana on the top of his head. It almost disguised his alien features.

Speaking of aliens, I started to notice that there were many different types of aliens in our larger group. Seth seemed to know them all; they would go off together to "play music". I wondered what they were really doing. A trip to the mother ship? Or maybe just a trip. LSD was the new recreational drug of the generation. This maverick never tried it. I was too scared.

I loved Seth. He had been my best friend, my protector, and boyfriend for many years, but there was still something about him that made me cautious. I often wondered if my caution was the influence from my parents interfering with the "free love" movement that I could never seem to join. Telepaths for parents—what a bummer.

"Flower for Peace, sir," I said to a businessman walking by. My now waist-length blonde hair had dulled to a mousy ash from lack of good shampooing. With my pendant rock and love beads covering the stains on my t-shirt, I looked like a poor waif, a beggar sitting on the sidewalk, legs crossed in a yoga position. I missed home. I looked sad, and the smile on my face seemed to be painted on, not felt. The grand mission seemed to have been better in my mind than in reality.

"Get a job, kid," the man said as he threw a quarter at me. It may have been that moment when I really began to wonder what on earth I was doing for the good of humanity. I hadn't had a decent shower in I can't remember when. I would have loved to roll in a bathtub full of lotion to help my parched skin. I would have died for the outfit of that girl who just walked past me as if I were an invisible ghost. I had to find Seth. We had to either complete the unclear mission, the destiny, or I would hitchhike home.

The pendant on my chest burned, and I looked up. Seth stood in front of me, his hand lowered towards me. I grabbed his hand, and without a moment's notice, we were standing in Time Square, New York City.

The air seemed to be sucked out of my lungs, and my skin felt cold. I remembered the darkness, the total lack of any light. I shivered. "So," I said in a timid, shaky voice, "this is weird, Seth. Where was the portal? You just touched my hand and here we are. How is that possible?" I was almost in tears at this point. This time dancing, as Mom called it, could have saved us a lot of time and torture. Why had we spent so much time in San Francisco and then just arrived in New York? Why were we in New York anyway?

"Anna, we're meeting the group from Berkley here in a few minute and then we will. Then we'll join them on their bus to Bethel, New York. We are part of the organizers for the three-day peace concert. We had to spend the summer in San Francisco to help gain numbers and be part of the recorded history, the Flower Power Movement. I know it was a long summer," Seth began.

He paused and then continued, "To answer your question about the portal, I really only needed the portal back to the ship on your mountain. This pendant that can fold time- you're wearing the twin," Seth said as he held the matching meteor rock towards the sky.

"This pendant has some limitations. You have to be able to visualize the exact location you wish to go and the exact time down to the millisecond. It's complicated. There is also the matter of controlling your unique frequency," Seth explained, his tone akin to a teacher addressing a class of kindergarteners.

What frequency? I had the twin to this magic timepiece, why hadn't I been told to use the dumb thing? Like Dorothy's ruby slippers, it seemed to have had the power to transport myself home all this time. I thought he could only vanish when going off-world to the Council chambers. That sucked. I could have been home each night, showered and well-rested, then just beam back to the miserable existence as a street waif passing out daisies. God, love a duck! I wasn't really feeling all that *"love, dove, peace"* stuff that was part of our mission to save humanity. What 'we' was he talking about?

"What 'we' are you talking about? You and me? Or are you talking about the kids from Berkley?" I almost yelled as frustration began to boil to the surface. Was 'we' those guys Seth slipped off with to "play music"? Was 'we' alien? My mind was spinning with a million different thoughts.

"No, 'we' as in the fellowship of the 'foreign exchange students', the junior observers. We're all here to watch history unfold and to be an active part of this remarkable event and help it along. We're also here to have fun. So get a real smile on your face. I have a friend that will let us stay the night. You can have your shower and your bath of lotion. Yes, I heard your thoughts. I understand," Seth said.

Suddenly, as if out of the blue, a really well-decorated "hippy" van drove up. The occupants rolled out while sweet smoke billowed out of the side door. They were very mellow and happy to see us.

"Seth," said the kid that looked a little like Big Bird, "we made it. Where is the safe house?"

Safe house, I thought, what are we now, part of the witness protection program or something? We all piled back in the van. The air was thick with the sweet smell I could only imagine as 'Mary Jane' or marijuana, if you insist.

The trip down an old Victorian section of the city was beautiful. It was early August, and the trees were just beginning to change from green to gold. The house was one of those old two-story Victorian homes on the National Historical Records. We walked up the brick sidewalk towards an ornately carved double door with a lion's head knocker. Seth knocked on the door as I admired the garden flowers.

The woman who answered the door looked like she just arrived off of a Bella Lugosi movie set. She had jet black hair down to her hips, white skin, and brilliant large oval green eyes. She swung both doors open and gestured us in, bowing her head in an odd fashion.

She ushered Seth and me up the ornate wooden staircase to a room on the left at the top of the stairs. She indicated that she expected us to occupy the same room.

As I walked through the door, I noticed that the room contained a wonderful claw foot tub and shower and a large feather bed. The floral antique spread invited me to jump in. I fell in love with the room, the sheets, the shower. I already knew that I loved Seth. I was just not ready to love him in that way... or was I? Or could this be our destiny maybe? I decided to wait and see if my father channels in before I let my virgin guard down.

Seth followed the woman back down the stairs. I showered until the skin on my hands began to shrivel up. Then I luxuriated in a milk bath for another hour. I felt so relaxed I must have simply fallen asleep. Somehow, I did not drown in the bath, but floated.

Morning broke through the lace curtains in our room. I noticed my naked body on top of the bedspread and toppled out of bed. Had I sleepwalked, or had Seth carried me to bed? I did have a dream about being lifted, but it was more like to the sky. It felt like flying.

If so, then I melted back into that big beautiful bed. I must have slipped off to sleep as time seemed to slip away, even though I had no memory of leaving the warmth of the milk bath. I was in heaven. An eternity passed.

Seth wasn't there. Had he ever been? Part of me felt disappointed and the other part relieved that I hadn't had to choose to lose my virginity that very night.

I got up and found a beautiful blue floral dress and sandals at the foot of the bed. I combed out my now-clean hair and braided it in one long French braid down my back. I put my pendant and beads around my neck and held the pendant in my hand.

Closing my eyes, I repeated, "There's no place like home, there's no place like home, there's no place like home." But I was still standing in that lovely Victorian bedroom. So much for the magic time dancer pendant. Maybe this was just the place I had envisioned in my mind anyway.

After a long look into the floor mirror, I made my way down the stairs to where I heard the voices of the group.

Warm blueberry muffins and hot tea were waiting for me, accompanied by a large tray of fresh fruit. I felt like I was in heaven and had no desire to leave. I hadn't heard from my parents for a while and began to wonder if I was somehow blocking their transmissions. The alien Berkley group, also known as the Junior Observers, had all gathered and were ready to depart. After eating, we started on our road trip to New York.

The bus arrived at Bethel. The crowds had already gathered. A hand-painted sign displayed the date, "August 15, 1969," followed by "Woodstock… Three Days of Peace." People had collected across the acres of land, in the mud, while the band played John Lennon's song "Give Peace a Chance". Thousands of voices singing in unison sent waves of vibrations throughout my body, and the pendant on my heart began to hum.

These children of the Earth, or perhaps the subculture of Flower Children, had indeed changed the course of history. Peace prevailed throughout the three-day concert and the one day of rest that followed. Never in the history of rock concerts had so many people braved the elements for so many days without a single harsh word or open conflict. This monumental event was a historical milestone. Woodstock was not just a moment in time, but a way of being in the world. This was how the future ought to be, with the children of the Earth singing, loving, and laughing in harmony. At least, these were the hopes and dreams of those who were part of the movement towards peace and love.

I realized it wasn't me or Wonder Woman who had created this history-making, world-saving event, but the children of Earth. This humbling thought made me feel incredibly honored to have been part of history, to watch history unfold. I marvel at the education of the exchange students on Earth, providing them with an active part in historical events.

Jeff paused for a moment thinking that it is how children learn about history. Through the folds of time, educators could transport students to historical events and allow them to witness history unfold in front of them. This privilege was only granted to the upper grades, who were allowed to participate in the events. Jeff stood up, stretched, and then Grandmother went into the kitchen. She said she'd prepare lunch and suggested we take a walk to unkink our bodies before continuing.

Lunch consisted mostly of fresh fruits and cheeses, with pita bread on the side. Both Jeff and I laughed at the strange revelations he'd uncovered after all these years living in a false existence. After downing tall glasses of tea, we walked across the red earth of Anna's lower yard and up to a hilltop with a vortex on the top. They sat on the hill and Jeff felt a surge of energy coursing through his body. He wondered if this might be another reason why his grandmother remained so remarkably youthful. Feeling refreshed, Jeff suggested we return to our activities.

The alien observers, the "band of seven," Seth and I remained at the farm and assisted the owner in cleaning the mess left behind by the more than 400,000 peaceful Earth children. We basked in the sun, slept in the barn, and waited for the time to approach for the big event in Washington D.C.

# CHAPTER 13

## Magical Mystery Tour

Finally, October arrived, and with it came the cold winds of early winter. The barn was becoming too cold to sleep in. Seth told me it was time to make our final journey. The wild Berkley bus filled to overflowing with Seth, the Junior Observers, the original gang of seven, and me.

A random thought struck Jeff. He wondered if the small paperback book, Snow White and the Seven Dwarfs, contained some coded message. He decided he would have to re-examine it again.

Our destination was the capital of the United States, Washington D.C., for what the world would come to know as the largest anti-war demonstration in the history of this country. Seth and I held our signs up, the ones we had created back in California. They bore the peace sign and the words "We want Peace, not War," painted in bright red. We marched with what was reported to be over 500,000 people.

We stood at the front gates of the White House in the evening, holding candles and singing anti-war songs. I marveled at the number of people in the crowd. My eyes moved around, looking at the candlelight and the dimly lit faces of peace marchers. Suddenly, I thought I saw someone that looked a little like my Mom, but my age, standing with the candle glow on her face.

"Yes, that was my mother," I thought to myself. "She had come to observe the events as they unfolded. She told me later that she had seen me there and was happy to see me warm and happy."

"Had your mother known of your adventures before you even had them?"

"Not everything, just those she was present to see."

Really, Anna, I thought to myself, your imagination is getting carried away. I guess I just missed my Mom. Suddenly a random thought crossed my mind. Had Mom really been here on this night, standing within eyeshot of what would become her daughter, me? I wondered. Maybe that is why she just seemed to know what to pack. I wrapped my arms around the warm wool poncho my mom suggested I pack and could feel her warm embrace as a soft memory of home. There was a surreal feeling that overwhelmed me. I needed reality. I needed home.

"Seth," I said as we finally made our way back to the bus, "was this the destiny you told me about years ago?"

"No, Anna, but it was part of the journey. There is another event that will change mankind's destiny. You will just have to wait."

"Whoa, now we're getting somewhere. Am I about to learn the big secret the Regents want to remain so secret? Am I about to learn the timeline anomaly?" I asked.

"Not quite yet," he replied.

"I thought you said this was our final journey. Are we on the way back home?" I asked. I was really ready to go home. I was homesick for my parents and even my little brother.

"Define 'home'," Seth said as he grabbed my hand and held his pendant in his other hand.

Suddenly, I was suddenly terrified. In a blink, I stood in front of a stone monument, a monolith. I was cold, breathless. At first, I thought we had somehow jumped to the Guatemalan Mayan Ruins, although the structures beyond the monolith looked very new, twinkling in the oddly dim iridescent light. The area was deep green with heavy foliage and oddly formed. As we walked towards the large stacked pyramid, I noticed very large orchids outlining the path towards the stone entrance. The air was heavy and the smell was unique, unlike any air I had breathed in.

"Where are we, Seth?" I asked, but I already had a feeling that we probably were not on Earth.

"This is my home, Anna. I have brought you to my people. You carry the hopes of the Galactic Federation as well as the Council of Worlds," he said, holding my hand and walking through the stone into the interior of the monument. Yes, I said "through," as if we had no mass, like ghostlike beings. My body just seemed to penetrate the solid stone entrance.

I attempted to pull away from Seth's grip as I stammered, "Kind of afraid here, Seth... So... what? Excuse me, but... WHAT? I'm carrying a WHAT? And how did we just do that?" I felt my whole body tremble.

Jeff thought that even in his reality, the ability to simply transport that amount of distance was still unknown science. This journal was getting more and more absurd, more abstract, more science fiction. But how could any of these historical events be imagined by my grandmother? How much of her journal was fact and how much was fantasy, he thought to himself.

By now, my voice seemed to echo inside the very stones we were walking through. I don't have anything but my poncho, dress, sandals, and my pendant. What exactly did Seth mean by that? The kind of answer I was getting in my head was one I was not willing to accept. What exactly happened in the Victorian Room in New York City? Isn't it just a little early to decide that I am carrying the hope of the Galaxy? Give me a break here, Seth. Seriously, dude, WHAT? I didn't think I was late. I don't really remember because of Woodstock, the barn, and Washington. It was all a blur.

"Don't panic, Anna. Everything will be fine. No harm will come to you physically or spiritually. I promise," Seth said. He walked up to what looked like white marble stairs which rose to a landing and another stone entrance. Again there was no visible opening, yet we passed through the stone into a large and well-decorated room. There were divans of solid gold and deep magenta tapestries laid on top. The white marble floor was cool to my feet, which were now bare. I suddenly realized that that wasn't all that was bare. I pulled back away from Seth. As he reached for me, his hand touched my middle finger. Just by that single touch to my finger, he made me feel comfortable, even in my nudity.

A beautiful woman appeared in front of me. Her hair touched the floor. The golden locks covered her body and became her robe. She had the same elongated skull as John and I figured it was his mother.

"Anna, I am Helena, the head of the Council of Worlds and emissary to the Galactic Federation. We welcome you to our home. This is Mira, a planet in the Tau Ceti Galaxy. The neighbor next door, as Seth has chosen to describe us."

Seth, one of our youngest observers, had reported to us that you carried Hope, who would join our world to yours. She was a blending of royal bloodlines that would usher in a new age of enlightenment for the "Planet of Children."

"What do you mean by 'I carry Hope'? Who? Seth and I have not engaged in inappropriate behaviors. We call our planet Earth, not the 'Planet of Children'," I said defensively. That unfiltered emotional outburst I had become famous for emerged at a really inappropriate time. My face must have been bright red. My whole body felt as though it had blushed, and I looked to the floor in shame.

When I looked down, I saw was nothing more than shocking. My flat stomach was now in full term pregnancy.

I think I must have fainted. Everything went dark and time stopped. I had a sense of time passing but no physical recognition that I was still even in my own body. It was more than dream-like, it was like a kind of death, ageless, timeless, pitch black, nothing.

When I woke up, I found myself in that beautiful Victorian room in New York. I was on top of the bedspread, naked. I was confused, to say the least. My hands ran down my naked body, no bulging belly. Man, that was a really bad dream. Seth wasn't there, just like I had "dreamed" before. I knew he would be downstairs waiting, or would he? I was stunned but thought to myself, it must have been one of those waking dreams when reality and dreams are confused.

The floral dress I had worn to the anti-war rally in New York and to Woodstock was neatly lying at the foot of the bed, alongside a nice pair of sandals I remembered wearing. Was this a dream, a vision of the future? Was this maybe the thoughts put into my head by my dad as a warning of things that might be, or had they already happened and I was time-danced back to my warmest memories for my sanity? Had my silly recital of "there's no place like home" brought me back here? Many questions rushed through my mind. What time was one reality? I really needed a clarification of time.

I got dressed and walked to the mirror to inspect myself. As I looked out the window, I noticed that the trees that had been just turning gold were now red and brown, and part of the leaves had fallen. I worried about the time problem but began to walk down the stairs to what was the dining room before the dream. The stairs began to morph into white marble steps as I descended. Terror overtook my whole body. I stood for a moment looking across the whiteness of the room. I saw Seth standing at the bottom of the landing. He held something in his arms. I was again afraid that it had not been a dream or a vision but a reality, a destiny that was playing out without my permission.

"Anna," Seth began, "I would like to present to you our daughter, the human hybrid, Hope. She will remain here with the Council of the Worlds and be taught the science of both galaxies. She will learn all the languages of the Universe and become the peacekeeper for the vastness of the dark and light energy of the cosmos." Seth was dressed in a solid gold toga with a chest plate that looked like something out of an old history book on Egypt. He held Hope in his outstretched arms and I reached to hold her in my arms. I still wasn't convinced I was really awake. I wanted to hold this child in my arms to see if she had warmth, mass, was flesh and blood, or just a vision in an otherworldly dream.

"Seth," Helena spoke sharply, "that is not a good idea. She will bond with the child and we can't have that. She will not be the mother to this child. Bring her to me."

Her voice startled me and I suddenly knew without a doubt that I was awake. That was my child and Seth was really standing there holding her in his outstretched arms.

Seth pulled the child close to his chest. "Helena, I promised Anna that no harm would come to her. You will not steal this child from her or from me. She carries the lineage of my forefathers as well as the royal genetics of her parents. She is ours. Anna, hold your pendant in your hand…NOW."

Seth had believed we would be with our daughter in Mira to help raise her and be a part of our life. It had never occurred to him that they would simply take possession of her as if he and I meant no more than slaves to them. Our young daughter was born on an alien planet in the galaxy of Tau Ceti on November 20, 1969. I held the pendant tightly in my hands and thought to myself, "There's no place like home." I could feel a deep cold envelop my body and darkness enclose around me.

# CHAPTER 14

## Bouncing the baby girl

"Before you say anything, I think I should explain," Anna began. "Seth and I had not had sexual relations even though it was a time of free love. We were both manipulated by the Council. His seed and my egg were manipulated with alien DNA and reintroduced to my body back in San Francisco. This is why I had that feeling of floating. They had transported me to a lab and the deed was done." She said it as though she was making excuses for being pregnant. Jeff rubbed her back and gave her a kiss on the cheek. He really didn't have words to express his confusion over everything.

> One day, I woke to the sound of a waterfall and the smell of fresh grass. As I rolled over, I saw Seth sitting on the grass with a small bundle between his legs. I moved towards him and saw a tiny baby wrapped in a gold fabric. Seth sat with the sun twinkling off the water droplets against his muscular, white nude chest. I sat down and the child opened her eyes. She had eyes like an angel, the brightest blue I had ever seen, beautiful and mesmerizing, just like her father's and golden hair like both of us. She seemed to smile.
>
> I lifted her into my arms and she snuggled against my chest. She was instantly mine. She was more than beautiful—perfect in every way. I nursed her for the first time and was so awestruck by the sensation. Tears formed in my eyes and Seth held me to his body.

"So, royal bloodlines? Can you explain that to me? And while you're at it, how did she happen?" I asked.

"New York and a bit of time dancing on the part of Helena. You had a natural birth even though you have no memory of it. Helena calls her Hope. Is that name okay for you?" Seth looked at me with more than a question, but a wonder as he asked.

"I think we should name her something else. She is part you, other-worldly, and part me, down to earth. So, how about Sky?"

Seth looked thoughtful and said, "Down to earth, land, sky, little bird, little fish. How about Skylar Hope Adamus? She is almost named after one of your great-great aunts, Skylla."

"That sounds wonderful, but how do you know I have an aunt named Skylla? I didn't even know that."

"You were kind of left out of the loop on your family history. You will have to ask your mother to fill you in one day."

"I can't wait to hear that explanation. I hope she doesn't dodge the issue at this late date. We do need to get Skylar a real birth certificate. So, where should we say she was born?"

"What about New York City, New York of the United States, and the good Earth in the Ra Galaxy? She was, after all, conceived there." Seth almost looked proud at that moment which made me a little unhappy.

"It's settled then. Now, where are we and where are we going?"

"I brought us to Salto de Chilasco, near Guatemala City in Central America. We have a little walk to the Candelaria Caves. I have friends within inner earth that will help us relocate for a few years while Helena looks for us across time and space." Seth cradled Skylar in his arms and we began to walk towards the mighty waterfall.

"The secret entrance to the cave system is under the falls, so we will all have a nice shower. Don't look at me with those big eyes, Anna, we are already damp. We will be able to send your parents a mental message as soon as we are underground, out of range of Helena's sensory reach."

We began to walk under the waterfall with the rugged rocks crumbling under our feet. I was so glad Seth had Skylar as I often felt my balance was unsteady. Once we entered the cave, the quartz crystals and golden colored rocks glistened in the dim light. In the distance, a tall alien-looking man came forward. Seth stood, and it appeared he was telepathically communicating with the man as I could hear no voices, just the sound of the running water.

We moved deeper into the cavern and the ancient ruins of a great underground city began to shine even in the darkness. We entered a columned doorway and the interior was decorated in ornate furnishings. It looked like a palace, a Mayan empire that had stood the sands of time and remained active throughout history. I then wondered if we were in our own time or that of the past.

Seth must have heard my musing and simply said, "Ancient history for you, yet now for them. You'll figure it out. This will be our home for the next year."

Seth and I became part of the community of the ancient Mayan. I am not talking about the Mayans in the archaeologist records. I am talking about the ones that took over the ruins after the Mayans returned to their own galaxy. The Mayan, the ancient aliens who built the place, kept the time logs of past and future events which would or could occur on Earth.

They were thinkers, mathematicians, scientists, astronomers, astronauts and I understood them perfectly. It was mind-boggling that I could understand their language as if it were my own, and it seemed unrealistic that I understood the higher mathematics and astronomy as though I were their equal. Seth had told me that my mind had been uploaded with the knowledge of the ancients. I was pretty impressed to say the least.

Jeff stopped and asked Anna where the crystal underground city was in the Mayan Empire. It had not been discovered in his version of history and it seemed like it might have been a part of the seven cities of Cibola. Anna said that they were indeed part of the underground world and that if he could continue his work, it might well be his next adventure.

Skylar learned how to read hieroglyphics and the ancient alien text. She also acquired knowledge in math and science, far beyond her normal earthly education. A typical earthling might have only learned to crawl, never learning to speak, read, or understand anything beyond survival needs. Skylar's aptitude was awe-inspiring. While Seth and the ancient Mayans taught Skylar, I learned about the great libraries of Universal Knowledge.

Seth told me they still exist in our time. The Mayans stored the stone tablets deep within the earth to preserve this knowledge for all time. I wasn't allowed to see or visit this marvel of ancient history, but Skylar was escorted by her father through the ancient, sacred texts. Seth informed me that the texts went back in time even further than their standards—to a time when the Reptilian race ruled the Earth.

I often wondered about the comments on the royal bloodline of earth and wondered if Skylar was somehow royalty, perhaps a princessif you will. By January 1972, our single year with the Mayan Empire had stretched into twoyears. Seth informed me that we would have to move. Helena had finally pinpointed the year in which we were hiding and had a general idea of what area of the planet we were on. Seth arranged for us to relocate to the city of Atlantis, which was actually a "mothership" sent to Earth's deepest waters, Lake Titicaca.

Seth reassured me, "Now don't get alarmed, but we are going to Atlantis in our own time. The ship was placed at the bottom of the lake because it connects to the massive caverns that join all the bases on Earth. From this place, we will be able to keep a watchful eye out for Helena...we have allies here."

We used the portal within the city of Chilam to transport onto the ship at the bottom of the lake. The city of Atlantis was unbelievably beautiful. Its walls were transparent, allowing the sea life to be seen, as well as letting its inhabitants be seen. The seabed seemed to be *simpatico* as though it were natural for this structure to be a part of the environment. I stood on the platform we arrived on and, I marveled at the sea life outside. Suddenly, I noticed a humanoid outside the enclosure, swimming with a great white shark. This creature was humanoid but altered—gills on a human face, long thin arms and legs with webbed fingers and toes. The humanoid had large eyes which blinked at me before it swam away.

Seth touched my arm, and I turned back towards the interior of the city. An Atlantian officer, an average-looking humanoid wearing a long white tunic and pants, met us on arrival. I wondered about the one I had seen outside the wall. Was this creature a different species from the one that welcomed us? He gestured for us to enter the long winding hall. The hall was made of marble, cream, and blue in color, and felt warm and inviting. We moved towards the center of the city. It did not look like a ship in the way I had envisioned a starship. The central plaza had a tall dome which looked more like a soft blue sky. When I noticed the stars moving and looked up, Seth simply told me in a one-word thought... fish. I laughed out loud, having already noticed the "fish" swimming outside the outer walls of the city.

Just as the knowledge of the ancient Mayans had been implanted in my mind, the language and knowledge of Atlantis seemed to come easily. As Skylar learned more about Earth than even the scientists knew, I learned about the science of the dome that kept the ocean from crashing through. I learned about plasma glass reinforced with carbon nanotubes and a third layer of photo chromatics. Because this beautiful dome was transparent and the aquatic world was so visible, it became my main study while on board the ship. I discovered that one of the wonders of this plasma shield was the ability to project a three-dimensional image on the outside of the dome to produce invisibility. I wondered if the Council could still detect us within the ship. I knew that Seth and I shared that same worry.

It was only a matter of time before our concerns reached the Atlantian Council. They not only picked up on our worry but also detected thought patterns indicating that the Council of Worlds might have found our location. The Atlantian Council came to a decision. We would be safer hiding in plain sight on the surface of Earth. We would be moving to another home, and they promised I would be reunited with my family.

# CHAPTER 15

## Time is in the moment

*Time dancing*, as my mother always nicknamed the act of moving through the folds of time, was a confusing event even on the best of days. The concept of one reality was difficult to master. Moving through the folds of time can only be done for viewing the events of the past. The future is still unwritten history, or so I was originally taught to believe.

Although my parents had traveled forward into the future, they told me it was because they hadn't altered the known events. Wasn't the birth of myself, my brother, and now the birth of the Universal Child, their grandchild, enough of a time ripple? Seth and I often wondered whether our decision to take our daughter away from Mira and the Council of Worlds was against the destiny which was foretold. We came to feel that the soul of kindness and the love we fought to achieve overruled the will of the Council or the Federation.

Before my brother graduated from high school, it was impossible for my parents to join us here. Now that my brother was off to the university, they sold the house at the base of My Mountain and moved in with us here in Arizona to help me navigate the time warp effect. It didn't hurt that they had a very cute granddaughter to help nurture either.

But what about the consequences of the time ripple we/they gave birth to? Just another worry, wonder, concern to occupy my mind.

My brother had gone off to the University of California at Berkeley. He had made friends with one of our original seven anti-war activists. They formed a friendship that would last a lifetime—alien and human debating the woes of the world, the galaxies, and the universe. Life seemed to be going on very well.

Mom and I stood at the window and watched the sun cascade across the red rocks of Sedona. It was the first light of a new day. Seth and my father were off scouting the vortexes for energy readings. They were monitoring who was in the area and whether we needed to relocate again in the now or the then.

Seth and I had traveled with our newborn daughter to all corners of the Earth. We jumped from Tau Ceti to Guatemala, down the bottom of the deep waters where Atlantis docked, and finally stopped in Arizona in the summer of 1973.

Jeff thought to himself, "So, this is when Grandmother first came to Sedona. I guess this is where she always felt at home. What would his life have been like if they had decided to make their home in El Paso? Would he still have graduated and got his masters? Would he still have had his wife and daughter? It seemed odd that taking one road or another road on the highway of destiny could change the course of history."

Skylar, whose name was supposed to have been Hope, was now going on 4 years old and already displaying rather disruptive telekinesis. There was no reason to wake her up from a nap like normal humans; she would just arrive on your lap when she woke. With her special gifts, it was difficult for us to make human friends and allow her to have playtime with the children in the neighborhood. I found that to be very sad. I had such great and strange memories of my time on Nations Ave.

The gang of cowboys turned FBI held fond memories for me. I had hoped that by removing Skylar from the clutches of the Council, she would be able to adapt to her alien half and blend in with the human population. Only time would be able to tell.

Skylar now had the knowledge of the ancients and the power of a mighty brain. Even though I wished she could form lasting memories with friends her own age, I wondered how that could ever be possible. She was too advanced to learn to play. What a shame.

I now understood why my parents had held my gifts back and not told me of my heritage. I was given a childhood, a normal albeit weird childhood. Skylar would stand out in the crowd of college students, not to mention preschoolers. It was almost sad that I couldn't find a way to provide the joy of early childhood to my own daughter.

I can tell that you have a burning question. Why, if you can travel in and out of time, don't you already have control over what may happen? The answer is simple. The one reality applies. You can visit history backwards in time, you can be a non-intrusive participant in time but the events that are destined to happen must happen, without too much interference or something absolutely terrible will occur—at least that's what I was told. Events in time can and have been slightly manipulated, but there is potential to create a rip in time which could alter all life in the cosmos. So we walk gently through time avoiding the ripple effect.

I asked that same question of my mother back when we finally talked about the escape from Titan and the Great Wars. It goes like this: Although you can travel in and out of time, time has full control of the reality. What is destined to happen will and must happen, the future is foretold but not written in stone. Destiny is nothing more than a roadmap for life, not a destination. It still sounded confusing to me.

Mom told me one of the weirdest and most alarming pieces of information in regards to that rip in time effect. When we had our abduction event back when I was twelve, my parents were told of the prophecy of Seth and I and about our unique and destined child Skylar. It had been written so it was destiny, not an anomaly which could cause a rip in time altering all life in the cosmos. Mom knew the prophecy of the Universal Child, but she never knew it was to have been Skylar. Mom only knew that Skylar had an important destiny, but not that their own grandchild was to be charged with the fate of humanoids within the cosmos. That was heavy pressure for sure.

The fact that they had informed us about Seth's and my destiny did explain why logically, morally grounded parents would have allowed their daughter to journey with an alien to San Francisco, and then New York. It was just destiny. I pondered what I would have done in this situation with my own daughter.

One of the greatest gifts is when you and your mother become friends and can speak freely about the past as well as the future. It was a time when I asked my parents those questions that had been plaguing me my whole life. The "Who am I?" and "What is my future?" questions, deep genealogical inquiries were nearly answered, along with what they knew to be future facts and what I could change.

My mother told me that part of children's education was to visit all habitable planets in the galaxy during historical events. The history lessons were designed to instruct in a very interactive way, to watch, observe, and learn from the mistakes made. The premise was that with knowledge of history, other wars and conflicts could be prevented from being repeated in the far future. I had never really understood why history seemed so important in school, but there was my answer. I don't think, from what I have witnessed, anyone on this planet ever really grasped it either. Probably, it is why we are destined to repeat the same mistakes over and over. Maybe that is why they call us the Planet of Children.

I often wondered if Seth and I hadn't already caused that ripple in time by removing our daughter, Skylar Hope Adamus, from Mira. Oh, by the way, that's her official birth certificate name. She was born in New York City, New York, United States of America, on planet Earth, not on planet Mira in the Tau Ceti Galaxy. That is the public record, not the reality of her true birthplace.

Seth told me he felt that although the Council wanted to control Hope, her real destiny was not to be their savior. He felt that her original destiny was to save Earth's destiny. I was still not convinced that she was to be the protector of Earth or the peacemaker for the Federation. To me, she was simply my very gifted little girl. Royal blood lines and all, our princess was to be a child first and a protector of humanoid life if she chose to, later.

Sedona was such a hotspot of UFO activity. To the Altantians, it seemed a perfect hiding place. The saying goes, if you want something to disappear, put it in plain sight. Seth and my parents agreed with the notion that hiding in plain sight was the safest. The people in the area were eccentric enough to accept strange people, events, or happenings on nearly a daily basis as normal. My nerdy self would fit right in. I hoped that as soon as my daughter could control her abilities, she could attend kindergarten here. Seth and I were homeschooling her, and her basic knowledge and capacity to retain information was beyond a twelfth-grade level. She would be more advanced than anyone in her class or even in all schools, but I wanted her to be able to make friends and learn about laughter and friendship from children her own biological age.

I had pondered what Skylar's mission and destiny would be down the line. She was maturing more each day and advancing in knowledge just as quickly. I often pondered the odd speed at which she had learned hieroglyphics and oceanography while still a baby and was understanding algebra at four years old. I wondered how much more her mind would absorb. She was like the library in Inner Earth, with the capacity to hold the knowledge of the Universe. Wow.

# CHAPTER 16

## Canyons beyond Time

The summer of 1975 was a time when the country was taking a deep breath after the fall of Saigon and the end of the Vietnam War. That war had brought Seth and me together as a couple and had driven our destiny. We had been part of the anti-war sentiments sweeping from coast to coast. Space flight dominated the news as the United States and the Soviet Union joined forces to send spacecraft to link up in space. Peace seemed to finally be at hand, and all seemed right with the world. Seth and I had a comfortable home in Sedona, Arizona. Skylar, a lively five-year-old, nearly six, was ready to take on the world.

Seth informed me that he had just received a message from his father, and he and Iris, Seth's mother, were planning to visit. Do you remember the movie "The Invasion of the Body Snatchers"? Having alien in-laws that you have never met arrange to visit is one of the most frightful and *horror show type* possibilities I had never even imagined. When I briefly met Helena of the Council of Worlds, I thought she was Seth's mother. I couldn't fathom what they would be like or look like, but they were coming in tomorrow night. The thing about time and aliens can be quite irritating, mostly at times like this.

My mother seemed excited to have them come, as if she already knew them. “Mom, aren’t you just a little bit worried about them visiting?” I asked, hoping she would let me in on the secret that she and Dad seemed to be keeping. It was always frustrating when they blocked me out.

“We were friends with them before the Great War," she replied. "Seth’s mother, Iris, as she is called today, and Horus were both cadets at the Science Academy when your father and I were there. We embarked on many adventures together. You know, we were even at the anti-war demonstration in Washington D.C. the same time you and Seth were there. I didn’t see you, but I was not supposed to be seen by anyone who would have recognized me. I wouldn’t have known you anyway. You weren’t even thought of yet.” She had a silly grin on her face and suddenly looked years younger. My mind flashed on the image burned into my memory, the face of my mother as young as I was, standing with candlelight glowing on her face.

“Mom… I saw you. I really did. I thought I was just missing you a lot. Wow…I still find time dancing more than strange."

Seth and my father returned from the vortexes. Dad told us that reports had confirmed that Seth’s parents were coming in by shuttle to the San Francisco Peak in Flagstaff tomorrow night. We could drive there and meet them, or they could simply beam directly here.

"Seth and I thought it would be nice to have a normal outing. It has been a while since we took Skylar out of the house," Dad said with excitement in his voice. I worried Skylar might lift the car off the road or do some other paranormal thing that "normal" Earth children couldn't do. It was true; we had sheltered her and kept her location secret, fearing the Council might find her and abduct her. We were probably being too careful. Maybe one of the other Junior Observers helped create a hybrid child. We hadn't heard from any of our friends. Maybe they were as guarded. We never told them about the birth of our daughter. Who knows, maybe she has friends to meet who are just like her.

Skylar was a prodigious girl who was still home schooled. Sending her to regular public education never seemed possible or practical. She had made friends with a few girls in the neighborhood and managed to hide her special talents. Skylar knew how to have a good time with girls her age, although, given her curriculum, she should have been in High School, perhaps even College. Skylar was approaching her seventh birthday, and her friends were the same age.

Although our home was rather secluded, it was still within bicycle range of the elementary school and the grocery store. Skylar had met a few children from the neighboring homes and always had fun with her new "human" friends. Although I always thought of her as my daughter and saw Seth as normal as apple pie, she was really an alien-human hybrid. Skylar quite liked her abilities and was sorry to have to hide them from her new friends. Unlike my memories, Skylar had a group of silly, laughing girlfriends who always wanted her to do sleepovers. She knew that would never be allowed. Skylar tended to sleep float, not sleepwalk. I always thought she looked so comfortable.

My mother told Skylar about the arrival of her other grandparents. She was so excited about meeting them that she literally floated above the floor. Skylar planned the menu and helped pick the fruit off the trees in our yard. She simply floated up to the tree tops to pick the best fruit. As she carried our picnic basket to the car, I hoped none of the neighbors noticed or drove by.

We waited for Seth's parents to arrive, got into our car, and off we went to the Grand Canyon. We sat on the edge of the canyon and had our lunch, talking about the history of the Annunaki and the other refugees of the Great Wars. It was reported that the Annunaki took up residence in the caverns deep inside the canyon walls. I didn't think that I had ever met any Annunaki, just Mayan and Atlantian inner-earth dwellers.

The Hopi and Zuni tribes, indigenous to the area, had a rich oral history here in the Southwest about the survival of the tribe. Tribal history tells of these refugees from the Great Wars, or star people as the Hopi call them, saving them before the biblical flood that wiped the Earth clean. According to the ancients, the relationship between the galaxies and the planet Earth went back billions of years, not just to the known cuneiform and hieroglyphic history, but well beyond creation.

Some information I learned after I first met Seth was disturbing. The Galactic Federation had the capability to cause extinction weather events through climate shifts. These anomalies, such as the Ice Age and the great floods, were a result of manipulating the flow direction of ocean water and adjusting the water temperature by regulating the radiation from the sun. I wondered if this method was frequently used as a sanction against Earth when we misbehaved. My fear of aliens, including in-law aliens, was based on these real facts.

We reached the base of San Francisco Peaks in Flagstaff, Arizona, around noon. The peak had an unusual glow to it, as if a cloud was resting right at the summit. Skylar looked up and said, "They're here." I wondered how she knew. Was the cloud hiding a large mother ship? I might be from an advanced gene pool, but some things felt too otherworldly for me to comprehend.

A California condor ascended above the canyon wall, soaring just above our heads. The afternoon sun highlighted their vast prehistoric wingspan and the air surged as they flew by. Skylar reached above her head as if asking them for a ride, and I held onto her ankles as she rose to meet them in mid-air.

"Remarkable," Iris remarked, looking at me as if I had interrupted a learning moment. "Anna, why didn't you let her go? They would have brought her back. She can communicate with them." She turned to my mother, who was peeling mangos and avocados. "Diana, have you not taught your daughter anything about our life, our abilities? She should also possess telepathy and telekinesis, right? She was born from your genetically advanced ovum and inherited abilities from Paulus."

"Iris, this is between Anna and Skylar. Your son and my daughter are both great parents. Maybe you don't know. Helena from the Council of Worlds wanted to keep Skylar at birth. They named her Hope. Seth and Anna had to defy the Council and hide from them. They've been living in fear of losing her since they escaped from Mira. Skylar has to blend into normal society for protection. There are eyes everywhere, and this is one of the most surveilled areas in the country. That's why Skylar is instructed, or rather, forbidden to demonstrate her abilities in public. The Grand Canyon is a National Park, extremely public and dangerous," Diana put down the fruit and walked towards the cliff edge.

Seth moved towards his mother-in-law and picked up Skylar in his arms. They stood motionless, facing the wind, and the sound of nature filled the silence, the tension.

Time seemed to stand still, maybe it was. The incident was set aside. The picnic commenced, and a familial atmosphere filled the air. Everything felt right with the world again.

History, one of the most significant legacies sentient beings carry into the future, is rife with mysteries, hidden agendas, conspiracies, wars, and conquest. With time folds facilitating life's motion, there were personal questions.

We had so much to discuss: our partnership with Tau Ceti through our union, the birth of our hybrid daughter, not to mention the comments made about altered DNA or my mother's genetically advanced ovum, resulting in me. Although I was supposedly born on Earth, there was that term: royal.

Oddly, my main wonder was the past relationship between our parents, the Science Academy on Titan, and whether they had orchestrated our union. Were we merely puppets in the hands of the Science Academy? Was I supposed to be Diana and Paulus' daughter, or was I just one of many sample experiments? I began to wonder about my brother. Was he also an experimental cell? We were both strange, nerdy, science geeks, and it somewhat made sense.

Were our parents and the Federation planning this union, or had Seth and I just randomly met? I knew my parents had spoken about the destiny revealed to them during our abduction decades ago, but in retrospect, I wondered. Seth's mind seemed to be on the same track. This was one of those unfortunate times when mind-reading was offline. Information wasn't being shared, and the gap between our reality and destiny had become vague and potentially perilous. Diana and Iris seemed to be transmitting some communications. I observed my mother's expressions fluctuating from laughter to terror to agreement. I wondered what they were contemplating. The car engine was the only audible sound.

The red rocks of Sedona came alive during our drive home. The brilliant colors created a picture-perfect landscape. The sunset blossomed into in full color. The red and golden rays of light made the saguaro cacti appear as obscure shadows like dark soldiers against the skyline, arms raised to the heavens, waiting to catch the last sacred light of Ra. The car was mostly silent, with everyone lost in their thoughts or sharing thoughts I couldn't read, except for Skylar. Out of the blue, she said, "Grandfather, did you know that magnetic forces align here? That's why it's so magical. My human friends tell me about something called 'enlightenment and meditation on the vortexes.' I've never been, Mom and Dad won't let me. Do you know why?"

Iris just mumbled under her breath. I sensed she thought I was keeping my daughter from advancing because I was too human. Horus, realizing that Skylar was addressing him, said, "Yes, child, it's true. That's why your parents keep you away. It's a very active portal area, and there are those who want to keep you off Earth."

"Why should they want to keep me off Earth, Pop?" Skylar questioned Paulus.

"The Council knows what a special girl you are, Skylar. You know you're my very special girl, don't you?" Horus piped in. "Your two grandfathers will always want you to be right where you're supposed to be, won't we, Paulus?" Diana started to voice her opinion, but Paulus gave her 'the look' - a facial expression that had become famous in our house. It was a silent message of "not now" or "hush."

There seemed to be a separation on political opinions between Iris and Horus about the Council's overall motivation towards Skylar seemed divided. My mother-in-law appeared to view Skylar similarly to how Helena had: as an untapped resource for negotiations with galaxies still at war deep within the universe. I believed Horus saw her role differently - to be here on this planet, in this galaxy, for the protection of humanoids within the Federation. I wondered if he might be affiliated with the Black League of the Orion Nebula, an historical alliance established as guardians of humanoids throughout the universe. This promised to be a very interesting, stressful, and eye-opening visit from the monster-in-laws.

Just in case you're lost, there are numerous galaxies within the known universe - more than 100,000 that Earth observatories in space have counted to date. Most are either involved in the Galactic Federation, or are too underdeveloped to join. However, there are those galaxies called the "Evil Six". These factions aim to commit genocide on all humanoid inhabitants of Earth, or Terra, and any other planet whose atmosphere supports humanoid life. Earth is particularly sacred to the Evil Six, who believe this planet to be their home world from trillions of years ago. Their mission is to reclaim and occupy their planet - our planet, Earth.

Their mission for Earth and all planets that harbor humanoids has not changed over the vast reaches of time. It's the same mission they had during the Great War. A band of six galaxies and the radical scientific minority on the Science Academy helped usher in the domino effect which ultimately created , leading to the explosion on Titan. This implosion of Titan, or Tiamat as it was once known, created a ripple effect causing loss of atmosphere to Mars and Earth's binary planet, the Moon. Advanced civilizations occupied these planets, and all were members of the Confederation of Planets. They had just enough advanced warning to create evacuation plans, searching for life-sustaining zones, places that could host humanoid life. This mass evacuation from Titan, Mars, Edam, and Earth caused many humanoids to become refugees among the vastness of space or retreat deep within the cores of the planets until the surface healed.

The Reptilian, Draconian, and Zeta Reticule from the Ursa Minor Dragon Council also had a seat on the

Galactic Federation Council but were never friends of the humanoids. The Ursa Minor ambassadors and the Orion Nebula councilors had deeply opposing views of dominion, victims, resistance, and connectivity. The political and racial separation of theology between the Evil Six and the humanoid members of the Federation had been the source of the Great War, which was still brewing in the outermost quadrant of the known universe.

This Universal political battle makes the petty arguing and war on Earth seem like schoolyard scraps in comparison. If Horus were indeed of the Black League, he would champion the idea that Earth should be left alone, quarantined, and declared off-limits by any faction aiming to change the natural course of evolution. I think that boundary was breached during World War Two when the Zetas provided advanced weaponry to the Nazi Party while making treaties with the American President.

The Zetas were said to have built a great underground city and military complex in the North Pole to use as an advance staging area, enabling the Nazi party to win the war. So why didn't they? Had the Federation intervened? Remember the Foo fighters seen in the air over Germany and the Pacific Islands? Were they aliens helping or hindering the outcome?

Did I forget to mention the transfer of information by Nazi scientists to America? The U.S. Government offered sanctuary to these scientists at Fort Bliss, in El Paso,

Texas, in exchange for advancements in rocket science and weaponry. I think the neighbor across the street was the wife of one of those scientists who briefly resided there. The secret city of Los Alamos, New Mexico, located between El Paso and Albuquerque, was that very city - secret and hidden down a long dirt road on the back side of a mountain, the same mountain range as the portal on my mountain. I felt a personal connection to this history.

Now, back to the subject of aliens and the monster-in-laws' visit. I had pondered the history of the galactic wars and the political differences among the planets and galaxies and began to wonder about Earth's mythology concerning the star people, of whom my love, Seth, was a part. Why were they visiting after so long? The Native Americans have an oral history of the visitors, or star people, who impregnate native women and then return on the anniversary of the child's sixth birthday to abduct them. I couldn't help but recall these mythologies and wonder why, after all these years, Seth's parents had arrived just short of Skylar's sixth birthday.

What role does this strange and complex merger between our two families was a component to the protection of Earth and the coalitions of humanoids within the Universe? Who are my parents, and who are Seth's? What does our daughter have that the Universe seems to covet? I was terrified at the thought. I began to wonder why Iris had insisted on coming to Earth to visit after so many years. Seth had reported the birth of our daughter to his parents in 1969; now it was 1975. Why now? With the ability to travel through time folds and with flying discs, it never takes long to get anywhere in the known universe. Why was this year so important? Was the myth of the Anasazi truth? Was I becoming paranoid and buying into a conspiracy theory? I needed to stop thinking negatively and imagine a positive outcome.

We arrived home and set up the table on the deck with beer and snacks. We showed them to their room. Then, I got Skylar ready for bed. The in-laws, my parents, and Seth sat out on the porch, watching the night sky. I excused myself and went to bed. It had been a very stressful day.

# CHAPTER 17

## Morning has broken

The early morning sun touched the tips of my eyelashes and I stretched to greet the new day. It had been a rather rough night, filled with thought-provoking dreams, sudden awakenings, and what I liked to call "sleep streaming. Dream dancing or sleep streaming are those moments during the dream state when you literally leave this world to adventure in the world of answers or longings, in the fourth dimension. I sometimes pondered the thought that moving into this dimension was like traveling to a parallel universe where the road less traveled could have been taken.

Jeff pondered that thought again, "the road less traveled". He wondered what the history of the world might have looked like if Gram had moved somewhere else.

I awoke with no answers, just questions. How could I play hostess to Seth's parents when I was concerned they, or should I say, one of them had a hidden agenda concerning my daughter? Seth must have awoken earlier and left the house. I got dressed and went into the kitchen.

I made hot red quinoa, an ancient grain from the Mayan empire, and set out a plate of fresh melon from the market with mission figs, and oranges from the trees in our yard. We had more than plenty after Skylar had topped the trees of all their fruit the day before. Seth and his father returned from their morning walk and were out on the deck overlooking the vast towers of red rocks. Iris was reading with Skylar in the den. I steeped a pot of green tea and called the family to the table. It all seemed so normal, so human, and so family-like, I almost forgot about the day before and my concerns. Maybe I had just become overly guarded, mistrusting and unreasonable. I would begin again with a brighter attitude this day.

"Breakfast, everyone, come to the table," I announced. I poured the tea and gave Skylar a glass of soy milk. I suddenly wondered how I had become so health-conscious. I used to love bacon and eggs and a hot cup of coffee with cream and sugar. How had I changed since Skylar was born? It was just weird, I thought. The 'do no harm' part of Seth's philosophy had led to becoming vegetarian, although I never truly gave up dairy products. Cows gave their milk freely after all, why should I deny myself ice cream or milk? But Skylar never liked anything that came from animals. Maybe she was more alien than human. Strike that…there are aliens that eat human flesh and animals as well. Just not the "civilized" ones, I guess. Were Earth humans not civilized? Was I truly a savage? I liked a good McDonald's hamburger and fries. My mind wandered and I suddenly thought, I must be hungry.

I sat down and the family filtered in and joined me at the table. My mom arrived from the side of the house and

I took a second look at her as she slid into a chair. She was holding something under the table.

"What do you have, Mom?" I asked as I grabbed a piece of cantaloupe off the plate.

"Just a rock. It looked interesting and has such an unusual luster to it," she replied. She held it out and Skylar screamed, rushing from the table. Horus took it and put it in a steel box that held recipes. "I think this has a highly negative magnetic quality to it. Maybe it somehow disturbs her vibrational field."

"Her what?" I asked, feeling as though they were talking about some other child, not my own daughter. Her vibrational what?

"Skylar has a very high vibrational field wrapped around her. I am surprised you can't see it for yourself, Anna. It is really very beautiful. Skylar literally twinkles. Sometimes, these children have that special gift; they cannot tolerate negative magnetic fields," Iris informed me with an 'I'm so superior' tone in her voice.

"What children with 'these gifts' are you talking about, Iris?" I asked, trying not to sound as completely offended as I truly was.

"Iris!" Diana yelled, "that will be enough!" Her face seemed to glow with anger and her eyes were like spears aiming directly towards Iris.

"Mom, let her finish. At this point, I think I have a right to a bit more information. She is MY daughter after all. What gifts and why do you see her glow? Why can't I?"

"I'm not really sure if you are ready for this information, Anna. This wasn't really the time to let the cat out of the bag so to speak," my dad, Paulus, said as though I were still in sixth grade.

"Anna, can we talk in the bedroom? You will excuse us, won't you?" Seth interjected as he grabbed my arm and gave both sets of parents a look I had never seen on his face before. It was a look of absolute fury. We entered the sanctity of our bedroom and I sat down on the end of the bed. I looked up at Seth's face and into those beautiful eyes and I suddenly noticed a strange look of fear, no, guilt, anger, maybe a combination, with a bit of concern that washed across his forehead and down to his now dewy blue eyes.

"Is there something you need to tell me, Seth?" I asked, still searching his face and his mind for clues.

"Our daughter is more than just a combination of Tau Ceti, Seth Adamus genes, and Anna Cronus's human DNA, but a blending of Alpha Centurion genomes. I am not exactly sure what other species were added after we arrived on Mira. Helena had you in her laboratory for many weeks. I really didn't think for even a moment that you, our child, or our life would be so completely under the control of the Council. I was young and gullible. There is no way I can even express my shame. Anna, I love you. I have loved you from the moment I brushed against your body back on Your Mountain fourteen years ago." Seth kneeled on the floor in front of me and placed his face into my hands. I felt his body shrink, his heart crush, and I wrapped my arms around him. We cried.

The moments seemed to last for hours, but the emotions of forgiveness were shattered by the now loud conversation which rolled like thunder through the canyon beyond our backyard.

"Why did you bring that damn rock in, Diana? You knew better!" Paulus yelled.

"I had to know for sure. Iris had threatened to expose us, to expose the Academy, the prophecy." Diana held her head in her hands and fell silent.

"She is almost six and it won't be long until she is of age. Time is fleeting. She has to be moved to the Council's control. She has to be educated. Her destiny is at hand. You have your emotional bonds connecting you to her as though she were a normal Earth Child. She is not. Helena was right. Anna never should have been allowed to bond with the child." Iris now screamed, and her voice sounded sinister, evil, threatening.

Horus was the only voice not heard. Was he still in the area? Was he on our side? Were my parents part of the conspiracy, or were they here to protect us and their granddaughter? Seth stood, his body seemed to become as glass, almost transparent, more than pale white, more than the color of thick cream, and I now noticed that his aura also seemed to be the color of an opal. I was in awe. He seemed to be in some kind of trance. His eyes were focused on the window out to the red rocks of Sedona.

I looked out the window and saw Horus. His eyes were deep purple and glowed like light behind a rich amethyst. His left hand was aimed towards the window with his palm exposed and an odd glow emanated from the center. Skylar was wrapped in his right arm like a football, and my heart literally stopped beating. Terror overtook my body, my soul. I felt as though I were paralyzed, a statue of human flesh standing next to a Greek sculpture. My world, my destiny, my best friend, Seth, frozen in time. I was stuck in motion and could not blink my eyes to shut out the terror of the scene. Then the world went black.

I woke shivering, gulping air like a guppy out of water. Seth was lying unconscious next to me on the warm grassy meadow. I looked at the sky and the snowcapped mountain beyond, and my eyes searched the meadow for Skylar or Horus. Where were we? Where was our daughter, and why was Seth still unconscious? I threw my arms around his body to see if I could feel his heart, his breath, his soul, his mind. I felt so totally alone and terrified.

Seth moaned and began to return to his normal creamy white skin. He sat up, and I watched as he looked at the sky, the mountain, and the meadow. His eyes were larger than I had ever seen. He searched every inch of the acres of flowers and grass. He closed his eyes and held the pendant close to his heart. I watched tears form in the corner of his eyes and roll down his Greek sculptured nose.

"What is it, Seth? Do you know where Skylar is?" I questioned with a quaver in my voice.

"The Law of One," Seth said with his head lowered to the ground. "She is with our fathers, but I am not sure in what reality." I was more than stunned, I was speechless. Tears began to roll down my cheeks and fall to the grass.

"Is that the same as the one reality?"

"No. I thought you understood what the one reality is. No matter when you are, you remain your chronological age. The Law of One is the law that states that all life vibrates, all life exists as one energy therefore you are able to be both solid and transparent, here and there all at once. You can travel between universes, or within the core of energy. In the short explanation, part of the One, the all, undetectable, untraceable."

"Can you explain that? Dumb it down or something, I don't understand."

"The world we live in is only one frequency, one vibrational reality of existence. Because all life vibrates, all frequencies, all expressions of life are the same energy. We are really each other. This is the Law of One. Moving or dancing through time occurs by changing one's frequencies or vibrations allowing solid bodies to change densities or dimensions. Wherever you are, whatever time you are in, you are existing in that single ONE reality, that I can find eventually. But the Law of One would be like finding a single grain of sand within the whole of everything existing everywhere. It's impossible." With that, Seth stood and stared at the sky and beyond.

The concept of the Law of One, the spirituality of the Universe, and the One reality all seemed to play major parts in the lives of those of us who live around universal destiny. Seth and I were pawns being played out on a great chess board somewhere in the known universe, with unknown agendas, pushing and pulling our lives like puppets marching to a silent band. Our young daughter seemed to be the queen that everyone wanted to capture, to possess. She was that separate grain of sand, the only solid in a fluid cosmos.

We were at a crossroads. Who could we trust? Even family seemed to have agendas, from my own mother to Seth's mother who seemed aligned with Helena of the Council of Worlds. We sat motionless on the warm grass meadow. We both considered our options or fate and the destiny we were, or our daughter was, to play in this new reality.

We began walking towards the snowcapped mountain. Seth had felt the pull of a powerful vortex and portal. We moved like zombies across the open land until we came upon the forest edge. There was a small rushing stream moving down the rocks, and we sat and drank until we felt our strength return. Seth held my hand and touched his pendant, and we arrived outside the lovely small town of Mount Shasta, California. We were just two strangers new in town.

We had to consider where we would live and how we could contact the fathers to find our daughter. It was fortunate that Mt. Shasta and Sedona had parallel personalities. The town had crystal shops and psychics, metaphysical shops, and tours of the vortex signs on buildings. I felt at home. We were odd, but the residents seemed to accept us anyway. People waved as though they knew us.

We stopped into a metaphysical candle and crystal shop and spoke to a lovely woman who was model tall with long, flowing, coal-black hair. She smiled at Seth as though they were old friends. I stood to the side, feeling invisible and even more alone than I had when I found myself in an unfamiliar landscape without my daughter and an unconscious mate. The woman told Seth about a welcoming center to which she would refer us. Seth seemed to be telecommunicating with her, but I could not hear her in my mind. It was as if my gifts had been stolen away with my daughter—perhaps with the arrival of Seth's parents. It felt as though I had simply gone deaf, at least within the range of my psychic abilities.

The tall woman seemed so trusting. She loaned us her Volkswagen bug and gave us the directions to the place she called the welcoming center. Seth and I found our way to a place called the Theosophical Society where Seth introduced himself as a mystic from Greece. We were warmly invited to remain at their campus. A small home was available for us to live in. I wondered what power the lady at the crystal shop had over these people, or whether she was an alien. It was entirely possible they were also aliens living among the humans.

Seth began giving classes on mystic enlightenment, and I worked in the library. I was treated a bit like an unintelligent child, so I cleaned the halls and straightened the gift shop. I took the time to read most of the books available in the general library on everything metaphysical. The people at the center were warm and welcoming but still made me feel out of the loop. It may have been my own impression, as my telepathic gifts were gone.

We waited for what seemed an eternity for word about Skylar. We missed her smiling face every moment of every day. Although I held a stoic face for Seth and the employees at the Theosophical Society, my heart was broken. I cried at night after Seth fell asleep, and I sent my loving thoughts into the vastness of space, hoping to hear something from Skylar. The months soon turned into a year, and the days, weeks, and months all seemed to drag endlessly by.

On September 19th, Seth came home in a panic, explaining that a mothership had been seen over Tehran, Iran, and another over Broad Haven in the United Kingdom. "It has been a decade, Anna. I have been out of the loop as an observer. Because we are in hiding, I wonder if the Federation has voted to sanction or remake Earth. Call your brother—he is living with Janus of Epsilon, an observer—they can tell us what is happening and if they have information about Skylar."

"Why would my brother know about Skylar and not tell us?" I asked.

"How would he have found us?" Seth countered.

We contacted Adam, and he told us that the climate change anomaly had been instituted. They thought the Confederation had slowed some of the most devastating effects, but that the Northeast would suffer extreme cold while the North Pole would be extremely warm. It was decided that it was not of major concern. Adam told me that all he knew was that Skylar was in a safe place on some sanctuary planet. It didn't make me feel any less unhappy. The Earth remained, and no extinction event occurred, just really odd weather. I was happy to know that Skylar was safe, but still yearned to find her.

November 20, 1976, arrived, and a deep depression filled my soul. Today was Skylar's birthday. She would turn seven today. Seth came home after his class, and we discussed the endless moments of not knowing where she was. Seth told me he had an idea of where we may be able to channel information. He had been able to access many of the documents in the secret library and knew of a location where hidden mysteries were stored.

It was still early enough to drive up the mountain and down the side roads that the forest service did not seem to maintain. We had been relieved to have been loaned a jeep for the trip, as our car would not have been able to come anywhere near this area. We came to the back side of Mt. Shasta and up a very rough trail. Seth and I got out and walked the additional quarter-mile on a deer track that seemed to ascend straight up.

An opening in the trees revealed a cave entrance covered with blackberries. Seth pushed aside the vegetation, and we went inside. The cave entrance grew into a large cavern as we followed petroglyphs deeper into the cavern. The Native American petroglyphs showed pictures of tall alien people, a story of water, sinking, and destruction of these people who came to this place to make a new home. They were pristine—I marveled at the thought that we may have been the first to set our eyes on them. No graffiti, no trash—I had to be the first human inside this cavern—ever.

The darkness in the cavern was beginning to cause me to hold on to Seth like a blind woman being led by a guide. Seth could always see in pitch darkness. We rounded a corner into another smaller cave entrance, and a soft iridescent light appeared. I could see the outline of what looked like a tall, thin person, or a starving bear. I was terrified and pressed my body almost into Seth. I heard a soft sound, almost like the sound of a crystal water glass with friction applied—bell-like, high, not really piercing but hypnotic, mesmerizing.

We continued to approach this being. I could feel my body pulling backward. I began to make out some features of this being. The figure was at least seven feet tall—maybe taller. He did stand a lot taller than Seth, who was almost seven feet tall himself. The figure had long, flowing hair that seemed to glow in the dim light. His face appeared to be transparent, slightly blue as though you could see inside the skin to the life blood. A long, earth-toned robe floated about. The creature spoke.

"Welcome to Telos. I am Meno, a Lemurian. What is your business with us, Seth and human female?" The creature gestured with a sweeping motion, as if to allow us to enter deeper into the cave.

"Why does he know your name and not mine, Seth?" I whispered, but my voice echoed anyway.

"You are here to ask about the Universal Child, are you not?" the creature queried.

"We are here to obtain news from the Confederation of Planets about the location of our daughter, Skylar, the one you call the Universal Child." Seth bowed his head as he spoke. I mirrored his actions, trying not to offend this strange cave dweller. I kept thinking...

Universal child... Why was she referred to in that manner? Had every species in the vast universe been told of her birth, and to what end? The mystery was more terrifying with each passing moment.

Jeff responded to this portion of the journal. "Gram, why don't I know anything about my older sister Skylar when the entire cosmos is familiar with her?"

Anna had tears in her eyes when she simply said, "Not now, that will have to wait."

Seth followed the odd creature deeper into the cavern. The iridescent light created a diamond-like glow on the interior walls. "I will guide you to the inner chamber. The Council Members have been told of your arrival."

This Lemurian walked ahead of us in the purple iridescent light. I could hear what could only be described as vibrational waves of bell-like tones coming from deep within the cavern interior. It reminded me of the quartz bowls used for vibrational healing at the center.

"Are they singing, Seth? What is that sound? Is that what their language sounds like?" I realized I had a bunch of questions, but Seth really could not answer me at that time. I guess my nerves and my compulsion to blurt out when scared or mad just popped out.

"Hush, Anna. All will be revealed in due time. Be calm," Seth reassured me. Although I was not calm nor could I be in such an alien environment, I tried to appear self-assured and purposeful.

I had read about Lemuria or Mu, as it was called in ancient texts in Atlantis, the two kingdoms which sank to the bottom of the ocean during the world extinction in the third age. It was thought that a comet or planet that impacted Earth created tidal waves and earth shifts, nuclear winter. It was called the age of Fire and Gravel. It was during this period of time the Annunaki saved the Hopi.

The Atlantians took ships to the heavens and to the deepest trenches in the ocean bottom. The Mu took shelter underground with the Annunaki and others. They used the deep caverns and volcanic tubes to move under the whole Earth at great speeds. I had always thought this to be mythology, not reality, but my life had been full of unreal events, and I was fully aware of the unknown secret life of the inner Earth. I also knew of the power of the alien scientists who could alter the weather patterns on Earth to cause extinction events. The inner earth dwellers had as much to lose as we above-ground earthlings.

We seemed to come to an end of the smaller tunnel, and ahead of us was a large carved stone wall. It had what looked like Egyptian figures and art I had seen on the tombs of the ancient pyramids. The Lemurian touched the wall, and it pivoted open. I saw a great hall with rich tapestry hung on the immensely high walls. The lighting inside was much brighter but had a foggy glow to the air. The Lemurian stepped aside, and Seth entered the hall. I began to follow, and this creature put an arm up to block my path.

The stone wall swung closed. I sunk to the floor. "Why can't I go in? She is my child... my daughter." The man just vaporized into the inner hall, leaving me alone and terrified.

Tears began to roll down my face, and I found myself helplessly gulping air. I could hear the bell tone rumbling coming from inside the great hall. I curled up into a tight ball on the cold stone floor and closed my eyes.

"Mom, I am here," I heard the soft voice of Skylar in my mind. It had seemed like an eternity since I had felt another's thoughts inside my mind that it startled me. I sat up in a flash.

Without thinking I verbalized, "Skylar, where are you? Are you okay? Are you with your grandfathers?" My voice came out loudly in the silent tunnel and echoed through the far reaches and depth of the endless underground world.

"Pop tells me we are on Chiron in the Alpha Centauri Galaxy. Grandfather tells me that they are arranging to have you and Dad meet me here. I think you are now with the Confederation of Planets Council, who are sympathetic to the mission of the Black League of Orion's Nebula. Grandfather just said that everything will be fine." There was a pause as though she were waiting for something and then said, "Pop told me to say, don't worry, and everything will be alright. You missed my birthday party but that's okay." A tear formed in the corner of my eye. Skylar had put this information in my mind, and although I was unhappy that I could not be in the council room with Seth and tell him what Skylar sent to me... I would be patient... I would sit and wait.

It seemed like an hour had passed when the stone wall pivoted open. The Lemurian motioned me inside the great hall. I rushed into Seth's arms. I didn't have a chance to open my mouth when Seth informed my mind that he knew Skylar had made contact. Again, I was struck with the surprise of his voice in my mind. Had my gifts returned suddenly?

Seth held my hand tightly, and I wondered what would happen next. This was usually when we time danced or moved through portals. I took a deep breath just in case. I held it until I realized we hadn't moved.

"Seth, what is happening?" I looked deeply into his eyes, trying to understand what was happening.

The tall Lemurian Councilor stood up and addressed me. "Earth woman, you and Seth will be reunited with the Universal Child. But I must warn you, she will have to be instructed in the laws and doctrine of the Galactic Federation. She is the hope of the universe, the peacemaker, the great moderator. You will have to be taught the history of your family as well as the history of the universal wars. It is important that you have full disclosure. It was not in your best interest to keep your genetic history a mystery. All of this will be explained in due time."

Seth still had a grip on my hand and I was beginning to wonder why. Suddenly, as if we had stepped onto a walking tram at the airport, we began to move, move faster, move even faster. The skin on my face seemed to be tucked behind my ears and the tunnel appeared to be rushing, swirling like fast moving water down a drain. I could still take a breath but I was paralyzed. I could not feel Seth's hand but I knew he was there. After a time, the tram seemed to slow and we stood at the opening of a cavern or opening somewhere. The air was dry, hot, more arid than Arizona. My eyes focused finally and I looked out upon a vast desert landscape. I could see the three pyramids of the Giza plateau in the distance.

Our Lemurian guide was standing behind us. "We will use the transporter platform in the king's chamber to transport to our ship. We have an observation ship on the dark side of the moon. From the ship, we will be able to transport you to Alpha Centauri and to your daughter. I have a report that your mother Diana will be joining you there."

"How will you be able to do that with all the tourists that visit the pyramids?" I just couldn't really wrap my mind around the logistics of this. Even with all I had experienced in my life with Seth, I still thought with a literal mind, a very human mind. Part of my mind wondered if we were to walk across what looked like miles to the pyramids across the hot desert floor. My face must have had questions on it. I was never very good at masking my emotions.

"I know that your education has been geared to a human child and not a star child, but you have the ability to vibrate at a higher frequency. This vibration is how we seem to move through solid objects, or through the folds of time. We can look invisible or ghost-like depending on the frequency. Seth will help you accomplish this task until you are able to do it alone."

Seth placed his hands on my shoulders and put his body close to mine. In this embarrassment, I felt complete. The dust and the sunlight made the air seem to become as diamonds against the sky. Suddenly, we were standing at the top of the Great Pyramid near the cover stone area where originally a cone of gold added extra power to the plant.

The pyramids were originally a worldwide grid providing power across the Earth by way of obelisks positioned around the globe. We stood in what originally served as a laser. It had been used to power the ships in the upper atmosphere. It was now used as a transporter unit. Seth stood with his arms rolled around my body, his hands still on my shoulders. The vast distance between the ground and our crow's nest was frightening and I was glad Seth held me so tightly.

We began to glow and again I felt the air become shallow and my skin begin to chill. We arrived on the Confederation of Planets vessel. A tall thin man in a white flowing robe approached us and guided us to our cabin. The room was stark but comfortable and did have a small portal in which to view the outside. I peered through the thick lens and saw Earth. It was truly breathtaking, like a gleaming blue pearl floating on the sea of sky. I hoped it would not be the last time I looked at my home world.

"Seth, how did we get from Mt. Shasta to Egypt?" I laid down on the bed and stretched my body. I had felt as though I was compressed into a small box from the moment I entered the chamber of the Lemurian Council.

"The Lemurian and Annunaki have used the volcanic tubes and underground caverns as highways around the world for millions of years. As a matter of fact, the Reptilian Races used these highways earlier than that. The Reptilian race are really one of the most dangerous species of Earth. They feel that Earth was their territory and humanoids are like an insect or virus that needs to be terminated. They are really pre-historic even by our standards." Seth sat beside me on the bed and rubbed my back. I had so much to think about, but I completely collapsed from stress and exhaustion. I soon fell sound asleep.

# CHAPTER 18

## Dark Energy

It was impossible to know whether it was night or day. I awoke to the artificial light of the cabin. Seth, dressed in a long white robe, was sitting on a chair by what appeared to be a small television screen.

"Morning, I think," I said to Seth, noticing how my voice seemed muffled, as though constrained by an invisible sound barrier.

"Good morning, Anna. There are clothes for you on the wardrobe. They expect us at the command center in a few moments. They've provided an energy shake for you in the silver vial. Bon appétit," Seth laughed.

I dressed in what resembled a long tunic and pants. The aqua fabric felt like air against my skin. I ran my hands down the tunic and marveled at its softness. It looked a little like an Asian pajama set, with slippers or booties for my feet. They felt like air,water or mist. I was reminded of the children's book, "The Emperor's New Clothes," and I wondered if the outfit was truly there. The aliens seemed to be a fan of nudity and were capable of telling your mind to believe something that was not there. "So, Seth, how does it look?" I asked, gauging his reaction. I then gulped down the vile and I do mean the vile mixture and made a face.

"You look like a student going to class. Let's hurry now," Seth replied, guiding me down the long hall. The interior looked like foil that had been subjected to a lot of heat; it bore an almost blue cast. Lights appeared from the dome of the corridor as we passed, like skylights without sunshine.

Upon arriving at the command center, I noticed the crew stayed focused on the task at hand. They were all dressed in tunics and pants, but not as loose fitting as mine. Maybe it was due to the fact they all towered over me in height and weight. Seth and a few officers wore white flowing robes, which made me wonder about his status. Seth had to have been given what looked like major officer's uniforms. The observation window revealed only vast, starless, black void.

A tall creature whose outer appearance was nearly transparent spoke, "We are within the space-time bubble traveling through the dark energy. The ship's bow diminishes the dark energy while expanding it at the stern. Since you're from California, you should understand the term 'surfing.' Propulsion through time and space is like surfing," the creature said in an instructional way. I got the impression he had 'dumbed it down' for me…lesson one…time bubble…got it. Seth and I were instructed to sit in what seemed like an observation row or more like a college lecture hall-type seating. We watched the officers and crew continue working the virtual screens as we moved motionless through time and space.

When the class or shift was over, we were directed to return to our cabin. When I asked Seth why he wore what looked like an officer's robe, he told me it was due to his heritage and that he had graduated from the academy many decades ago. Again, I wondered how old Seth truly was. The question of his heritage would have to wait for a better time. I knew this was not the time or place for a long conversation on genealogy.

Every day that is just what I called it because there hadn't been anything to indicate a change in night and day; it just seemed to continue on. There was a time for nourishment (not food, mind you), a time for rest and a time to sleep. The lack of sunlight made me feel like I was suffocating. It was unnerving. The total lack of color in the monotone ship was dreadful. I remember counting the times we went to our cabin to sleep; nine nights had passed. On the tenth day, we woke to the sound of alarms, bell tones and communication I could not understand. Seth quickly dressed, throwing my 'uniform' at me and pulling me out the door half-dressed.

"We have arrived at our destination," Seth announced, practically running towards the observation room. As I struggled to don my booties on hopped behind him like a wounded bunny. The door slid open, revealing a light so bright it hurt my eyes. It was a stark contrast to the half-light we had been living in for ten days. I closed my eyes tightly and opened them again. Seth pressed a button on a control panel, adding a protective covering to the window to make viewing more tolerable. It must have been that same technology the Altantians had in the dome above the main plaza, the photo chromatics layer.

Outside the observation window was a metropolis made of clear quartz crystals under a deep purple sky tinged with magenta. It was beautiful. I could not tell if it were early morning or late evening, but the sky reminded me of a sunset in Sedona with more purple than I had ever seen. The city reflected the ambient light, creating a soft prism-like glow on the sharp edges of the structures.

"Are we here? Is this where Skylar is? What planet is this and what galaxy?" I could hardly contain myself. My concerns about the planet's suitability for humans were pushed aside by its sheer alien beauty. Before Seth had a chance to figure out where we were in the cosmos, a tall, white-skinned humanoid alien with enormous blue almond-shaped eyes was dressed in a black skin tight uniform, directed us to a holding room. The alien pointed towards the far wall. With its fingers outstretched, I noticed it only had three digits on each hand—*Arcturian,* I thought. The door sealed behind us leaving us alone in the holding room. It was stark, devoid of color or windows or portals, so white it suffocated my mind and soul. Seth turned me to face him.

His face always had a calming effect on me. How could he always soothe my soul without a single word or emotion?

"Anna, close your eyes tightly and don't open them until I tell you to. Do you understand?"

I closed my eyes. The act of closing out the insufferable emptiness of the white holding room eased my tension. "Seth, this is new. Should I be worried? Should I hold my breath or what?"

Before he had an opportunity to answer, I heard what sounded like crackling static or electric shock waves. I felt a tingling everywhere, and I thought my hair must have been standing straight up. It felt as if an electric charge had frizzed the hair on my entire body. I stood there, eyes closed, trembling and afraid. Had we warped through time or space again? What was this terrible electrical charge I felt flowing through every muscle, pore, and blood vessel in my body? My nerves were tingling. For a brief moment, I had a sense that I was almost weightless.

"It's over, Anna. You can open your eyes now."

I looked around, expecting to still be in the holding room, but we were on what looked like a polished silver floor in an entirely glass or crystal room. The doorway was open to the plaza in what I assumed to be the metropolis I had seen from the observation room. "Seth, do you know where we are?" I ran my hands down my hair to see if it was lying flat or still sky-bound.

"I think we are on Chiron, a planet in the Alpha Centauri system. I am just guessing because of the crystal structures. This planet is about the same mass as Earth, revolving around the red sun—one of three in the galaxy. We should be met by one of the guides soon. Be patient."

I looked at the sky and noticed a transparent dome like on Atlantis, but the sky appeared to be in unnatural shades of magenta and purple. This could have been the color of the dome changing the night sky, or maybe it was the color of the sun. I could make out in the distant sky a red orb and a smaller orange orb—a moon, or was it the sun? In the deep purple sky, I could make out star constellations, but I did not recognize any of them. From that painfully colorless room to this all-encompassing array of majestic colors, I was awestruck. It was beautiful, alien, and abstract.

Just as Seth had told me about the guides, three uniformed humans approached us. They had insignia that read Sanctuary Patrol. Before I could ask Seth what a Sanctuary Patrol Officer was, the tallest male spoke.

"Sir, you are from Mira of Tau Ceti, but your Earth companion needs her papers. Do you have them?" he asked as he took my elbow as though I had been arrested.

"I am Seth Adamus, and this is my mate Anna. We are here by way of the Confederation of Planets to find our daughter Skylar, the Universal Child. Does Anna still need to have papers?"

"No, sir, sorry for the confusion. Welcome to Chiron. We are considered a sanctuary planet for humans who ask for protection. Earth humans need pre-admission papers from the immigration teams on Earth to be able to seek sanctuary from the secret government on Earth. We have to check papers. There are traitors to the cause and spies. We must be cautious."

"Do you know of our daughter?" I asked the tall, brown-skinned man.

"I have heard the prophecies and was told she was here in protective custody."

I was alarmed by his comment. I did not attempt to get any more information from the officers. I had noticed that the female officer was Asian, with black hair, almond eyes, and white skin. The other male officer was black, with round, black eyes, looking like a police officer from Phoenix. The people on this planet were human and just as diverse as Earth. Other than the odd architecture and strange color of the sky, I would have thought I was still on Earth.

We walked down the silver cobblestone-like sidewalk with the fountains of water flowing around the multi-colored gem sculptures. The potted plants were odd, with succulent-style leaves of red, pink, and pale veins. The flora was unique but beautiful. The entire ambiance was artistic, abstract, and breathtaking. Soon, we stepped on another tram-type walkway. I grabbed Seth's arm, fearing we would take off like we had in the caverns under the earth at home. Seth smiled, and we traveled at a comfortable speed past children playing and people sitting as though they were in a park on a Sunday afternoon. The brown-skinned officer took a step to the side onto yet another walkway, and we followed. He stepped off the path onto a courtyard of rose-colored crystal gravel and motioned us towards the clear steps.

"Just up the stairs to the left, you will find your accommodations. Your daughter and the other members of your family have been advised of your arrival. Good day to you both."

I practically flew up the stairs and to the door on the left. I could see inside what must have been the main room, and I took a step back. Seth rushed up to meet me and looked inside. Skylar was floating just off the silver floor, completely surrounded by four-foot-tall columns of lapis lazuli crystals. There was a circular bubble of opal light surrounding her. I looked at Seth in a panic. "What is this, Seth?"

"I think she is being assisted by the power of the lapis to activate her higher mind and her spiritual enlightenment. I have to admit, I can't ever remember anyone that young to have been placed in the light body before." Seth's face beamed with pride, but I was alarmed, not proud. I was concerned and confused.

Seth placed his palm against the door, and it vanished. Skylar looked up from her meditation and dropped to the floor. She flew into her father's arms and then to mine. I held her so tightly she began to complain. As I set her down, I saw my father come out of a side hallway followed by my mother and Seth's father, Horus. I held my breath for a moment, wondering if Iris would soon follow, but she seemed to be absent. I must have let out an audible sigh of relief. I could feel Seth inside my mind, assuring me that Iris was nowhere on this planet. Everyone began to talk at the same time, and none of us could understand what the other said. It didn't really matter. We were reunited. The family was whole. Home was wherever we were all together. We were home.

I walked through the apartment, inspecting each room. The furniture resembled a designer loft from New York City—ultra-modern yet adorned in rich colors of blue and mauve against stark white furniture. Crystal tables seemed to contain the light as though they were lamps. Odd house plants decorated the corners. The bathroom in what I was told was our room housed a large shower and soaking pool made of clear crystal. It looked inviting, although I noticed a plethora of large mirrors everywhere. I muddled over the thought that even alien architects must be vain.

There was no kitchen that I could recognize, but an eating area existed with yet another enormous mirror and what appeared to be a large television set. No artwork adorned any wall, just mirrors. A silver and copper-colored table was surrounded by eight chairs. There were no cooking appliances or cabinets. "Must be room service all the time," I thought. I returned to the main room to look for everyone. The room was empty except for towers of lapis lazuli. I stood motionless, listening.

Suddenly, I heard Skylar's laughter coming from a room just to the side of the front entrance. I had walked to the right and all through the apartment but had neglected to notice an entrance to the left. Perhaps this was because any wall could become a door in this place. I entered through an arch to find a large living room with plush white couches and easy chairs. The palette was in shades of purple, cobalt, and cream. The whole family was seated around a pure white rug on the silver floor. Skylar was lying prone on the rug, elbows holding her chin, golden locks of hair flowing down her back. She was the center of attention and knew it. That image was so beautiful I instantly filed it in my memory as though this would or could be the last time I viewed my little girl. The shock over her abduction and the never-ending search for her were fresh in my mind. Perhaps I did not see this image as reality but part of an extravagant nightmare.

"Mom, I thought you would never get here," Skylar yelled as she rushed into my arms. Her voice brought me back to the present.

"Time is a funny thing around here," Skylar started, responding to my unspoken question. "I can't remember how long it has been since we were at home in Sedona. Grandfather and Pop have been making sure I do my homework. Grandma has been watching the Council teachers so that they did not take advantage of my time. She REALLY watches them too—I think she makes them more nervous than they make me." With a quick kiss on my cheek, she slid out of my arms and onto the floor, rushing towards Seth and landing in his lap. They both laughed.

A random thought crossed my mind: why did she run, not float? Council teachers? How could the Council have anything to do with Skylar? Chiron was a sanctuary planet under the protection of the Confederation Council and the Black Orion League. I didn't think the Confederation had teachers and mentors outside of Mt. Shasta. Was Chiron a sanctuary planet or a comfortable prison?

"I think it's time for dinner. I feel hungry, don't you?" Mom said as she walked through the main room, or Skylar's homework room, and into the television room with the table.

I followed her in. "So, where is the fridge, the stove, the sink, and maybe some dishes?" I asked as I sat in the side chair of the silver and copper table. I considered within my mind, No cooking on Chiron, just room service here. Just like a prison, the jailers bring your food and slid it under the door, my mind wandered into the conspiracy theory that had always been part of my personality.

"Just being silly," I thought to myself, but I was still worried.

"George Jetson-style kitchen, honey. Just tell me what you want for dinner, and it will be ready in a flash," Mom said, waiting for the rest of the family to pile into the eating area. After everyone sat down, she touched the television screen, but nothing seemed to happen. I didn't know what I thought would happen, but nothing changed in the room. "Interesting," as Spock would have said on that television show, Star Trek.

Skylar blurted out her order, "I want a big grilled mushroom with tomatoes and avocados on a hamburger bun and fries—oh, and I want a really big soy shake with chocolate too."

"I'll have a porterhouse steak with a tossed salad and onion rings, and a glass of Merlo wine," Horus said. Seth agreed to have the same thing.

I wondered if the steak was made of animal flesh or if it was a chemical composite. After all, life was sacred for most of the humanoid galaxies, with the exception of some Earthlings who loved beef hamburgers and fries.

"So, Anna, what's your poison tonight?" Mom asked as she patted me on the back. "Come on now, it's all about what your mind really wants. Give it a try."

"I don't think I am hungry. You all go on, I can't decide." I really couldn't think of a single thing to eat after more than a week of liquid nourishment. Solid food sounded dreadful.

My family was sitting around a table, together like we used to do back home, without Horus, of course, but it felt like home, even though I had not been able to prepare a meal for my family. It felt really strange to be sitting around a dinner table, looking at an empty table. Just weird!

The meals and drinks seemed to just appear on the table in front of the person who ordered. I looked down and I had a bowl of San Francisco-style clam chowder in a sourdough bowl and a tall pilsner beer. I really had thought about that soup for more than a week while on the ship. So, was this like Peter Pan, we had imaginary food? It all smelled wonderful and I began to taste the soup—it was warm and full-bodied. The heavy liquid rolled down my throat. It might have been an illusion, but my senses were in heaven.

The big screen television, or monitor, seemed to vanish, and the large mirror suddenly became a picture window. Out of that virtual window, the red rocks of Sedona appeared, including a sunset of gold and magenta. The wind blew the trees as though I were looking out the patio door back home. As we sat eating our dinner and commenting on the taste of the food or the illusion of food, the sky out the virtual window darkened to night and the stars of Earth appeared as though we had been transported back to our home.

"So, is this virtual window the same in every apartment around here?" I asked.

"Actually, the system chose this for you. Yesterday, the window reflected the skies of Titan. It's custom like the menu, made of thoughts and emotions of the occupant. You must have had the strongest images in your mind." Mom just blurted it out as if I should have figured it out for myself.

"Why Titan yesterday and Sedona today?" I asked, curious.

"You'll figure it out in the next few days," Seth patted my arm. It felt like he was telling a small child, 'not now'. That made me a bit irritated, but I was still so happy to be together as a family, I let it go.

When we had finished dinner, Skylar yawned and said, "Mom, can you put me to bed? I really missed that." She walked in front of me into her room, which was lit by the crystal tables beside her bed. I hadn't noticed her room originally; I think I was so overwhelmed by the practice room and Skylar floating in the air I never thought to see where she called home.

Her bedroom was warm and appealing and included a chest with her toys that must have been reproduced from her memories. She handed me a book, Snow White and the Seven Dwarves, and asked me to read it to her. The last time I remember reading her this book was when she was two years old. I wondered why she found this old book so meaningful. The image of my Evil Grandmother, Iris, suddenly came to mind and I smiled as I began to read. The image of the seven dwarves now looked more like the towering alien junior observers we had traveled with during our trip to New York, only in my mind I was the only dwarf.

As I read the story, my mind rewrote the simple fairy tale and I began to feel myself yawn. Skylar had never been told the stories of the seven travelers, but maybe cellular memory was a factor. Skylar slipped off to sleep. I stretched and walked to the door. I would have turned the light off at home but there were no switches. As I looked at my precious little girl, the lights dimmed to candle light strength. The virtual window shades closed, and I wondered if that mirror/window wasn't also an observation screen. Was someone watching?

I returned to the living room and sat next to my mother. I needed to ask her a lot of questions, and I wanted to know whose side she was on. I curled up on the couch and pulled a cobalt cover over my legs. The living room window showed the silhouette of Bell Rock and the night sky behind. Again, who was watching behind that curtain of virtual imagery?

"You have questions, Anna. I just want you to understand that I am not on the side of the Council, but I did think it very important that Skylar be given the opportunity to learn and to be able to choose to fulfill her destiny. Iris just wanted her separated from humanity. I really never could understand her detached feeling over her own granddaughter, although it may be a species thing when I think about it." Mom rolled her arm around me and pulled me towards her.

"Mom, I don't understand what the destiny or prophecy says about Seth and me, or about Skylar. Is there something I should be told or is there something I could read that would help me understand? And, species thing, are you talking about aliens like Tau Ceti?" I lowered my voice to an almost whisper, trying not to glance at the window in fear someone would know I figured out the surveillance.

"No, not Tau Ceti. That will have to be a conversation we have at the Archival Data Retrieval Center later. My fault, Iris was right about one thing. I should have told you everything about you and your brother. I should have informed you about the contract we signed with the science council before you were even born. And I should have taught you the history of the Universal War, the destruction of Titan, and the prophecy you, Seth, and Skylar were to fulfill. Can we just take this moment in time to be just human, just mother and daughter, and think about the past and the future tomorrow?"

Mom buried her eyes in the cobalt cover, pressing her head into my lap. I became pretty sure she knew someone was monitoring everything within this home. I turned my back to the window and patted mom's head. I whispered softly, as though I would wake up a sleeping giant, "Tomorrow then, but you can't leave anything untold. No more secrets, do you understand?"

I didn't let her know that I had already researched the Great War or the Universal War as she had called it. I had already studied the destruction of Titan as well as the exodus of the humanoids as they scattered across the cosmos in the library of Mu. I studied what the Mayans let me access when we lived with them and knew that they were also shielding me from some information about genealogy, my little family, and our combined heritage. I wondered what this top secret, eyes only truth was.

My mind was tumbling with questions and a desire to know the truth. Seth couldn't, wouldn't reveal anything to me. Man, talk about living on the fringe of your own life. I would just let Mom tell me in her own words in the morning and use my intuition to determine if I believed her or not. I was still curious about my genealogy and the contract my parents had signed. Seth and Skylar being referred to as royal bloodlines; it was as though I were an adopted kid without a family history to fall back on. Maybe that was the real truth; I was abducted as an infant from regular earth parents, just raised by what I thought were my parents. Stupid conspiracy theory, you look like your mother, I thought, and put it out of my mind, for now anyway.

As I excused myself for the night, my mind continued to mull over the thousand possibilities that could have resulted in my family's refusal to tell me the truth. Seth joined me in our room. This room was as warm and inviting as Skylar's had been. I touched the wall Seth had shown me and retrieved a nightgown and robe.

Behind the wall, separating the bath and our room, I disrobed and slipped into the crystal tub, which was already filled with warm, inviting water. I sat there, relaxing, until I began to feel myself slipping off to sleep. Seth touched my shoulders and told me to come to bed. The window, or was it a monitor, reflected the night sky and Earth's moon high above. I pulled the robe tightly around my body and slipped into bed. Automatically, the shades closed. Silently, I thought but did not verbally ask, *Who is watching us, Seth?*

"This is an autonomous dwelling which reads the individual's thoughts and body vibrations," Seth explained. "The window is there to bring mental health to the occupant by reflecting the tenant's most comfortable images of places they feel at home. There is no biological being monitoring us. Sleep well, and we will talk in the morning."

Seth pulled the covers over his beautiful nude body and gently kissed me goodnight. I pulled off the robe, dropped it to the floor, and rolled into Seth's arm. He was my private sanctuary, my physical home, my soul mate. The heaviness overtook my desire, and before I knew it, I was gone, deeply and soundly asleep.

# CHAPTER 19

## Here comes the Suns

The morning sun broke through the window in our room. I covered my eyes for a moment, allowing them to adjust to the morning light. The sky was light blue, and the wind was gently blowing the leaves of the orange tree outside our bedroom window. I heard the chirping of the house finch that woke me most mornings in Sedona.

The bedroom wall shone brightly as the yellow morning sun of Earth rose across the red rocks of Sedona, Arizona. For just a moment, I thought I had simply awoken from a terrible nightmare of abduction and travel across the cosmos. I looked out the virtual window and watched the small house finch sit on the orange tree pruning his feathers. I felt more rested than I had in years. Seth rolled over and kissed me. For a moment, I had traveled back in time.

"Are we home?" I asked as I rolled my arms around his body.

"Define home," Seth said as he pulled me to his body.

For just a flash, I thought of standing in the street in Washington D.C., just seconds before transporting to Tau Ceti…define home… "Not again, Seth…are we still off world? Are we in the Tau Ceti Galaxy or are we in Alpha Centauri?"

"Boy, you really did sleep like the dead last night. No, we are not on Tau Ceti, and yes, everything that you remember really did happen. We are on Chiron in Alpha Centauri. Now come here." Seth's warm muscular torso embraced my whole body, and for a moment, I melted into his arms. As I felt my body heat rise, I suddenly stopped. My mind went from pure passion to an overwhelming motherly instinct. I pulled away, and Seth looked wounded.

"Not now, Seth…I need to see Skylar with my own eyes this morning. Let's go find her." I pulled the magenta cover off and pulled the robe around my body. The thought that someone was watching was still very prevalent in my mind.

"Don't be so suspicious, Anna, get dressed," Seth stated, almost like a parent, not a partner, as he slid open a wall area, and a closet full of nice clothes appeared. I disrobed and picked a pair of jeans and a t-shirt from the hanger. The fabric again felt like air, and I still wondered how much was real or planted in my mind, just like the holographic sky of Earth outside the mirrored window. Was this fabric real or imagined?

We went into the kitchen and found my parents having a cup of coffee. It seemed wonderful.

"Where is Skylar?" I asked as I sat at the table. A hot cup of coffee with cream and sugar manifested in front of me.

"She has gone to school," Diana reached for the center of the table, and a bowl of Earth fruit appeared.

"School? Skylar has never gone to school; she has always been homeschooled. Why now…why here, Mom?"

"Skylar is with earthlings or should I say hybrid children of Earth. She is in a class of her peers. She does have special classes she attends after school with the Arcturian Elders who are teaching her how to use her gifts and to become all she was born to be. She has been able to use her innate vibrational ability to disappear. I'm just telling you this so that she won't scare the heck out of you." Mom started to laugh. "The day she learned how to master this, she came home from school and did one of those ghostbuster-style howls behind me, and I about jumped through the ceiling."

"How long has this going-to-school thing been going on?" I whispered, still worried the information was somehow top secret. My mind could imagine a group of aliens sitting at a long countertop watching me from that virtual picture window. It felt unnerving and uncomfortable.

"Skylar started classes after she arrived here. We had to convince her you knew where she was and that everything was okay."

"Well, it wasn't okay, Mom, and we had no idea where she was. Do you have any idea how much torture these people have put us through? Really, Mom, you sound as if this isn't a big deal. It is…it was...and I don't think it will ever be okay…do you understand?" I couldn't help wonder why she did not know that it was Skylar and Dad who contacted me telepathically in inner earth. Skylar knew I didn't know where she was…really. I almost came to the point of yelling as my eyes darted all around the house for monitors or cameras or other "Big Brother"-type devices.

"Anna, I know that the last time you saw Skylar was when you were at home in Sedona. This protective service your father and father-in-law perpetrated, what you thought of as abduction, was necessary. I was the distraction. Horus had to subdue Seth…his powers are much greater than his father's. We were all so sorry it had to be done this way, honey. We sent you to the only place we could think of that we thought you could find your way back to us. We had learned from a secret correspondence that Iris was planning on stealing Skylar and taking her to Serpo, a Reptilian, and Zeta stronghold."

"Serpo? Is that a constellation?" I was getting confused.

"It is in the Virgo Constellation and is rather like Chiron, only a base for the evil six, but this is off topic, honey. Helena, even though she is an ambassador of Tau-Ceti, is also a high-ranking politician from the Reptilian

Alliance. Helena attended some of the genome classes at the Academy. It was on Titan that Iris and Helena became friends."

"Is this why Iris decided to steal Skylar? Was it the influence of Helena that made her jeopardize her own granddaughter…or was it a species thing…whatever that is, Mom?"

"Iris joined the Council of Worlds as an ambassador of Middle Egypt. It was during the third age the split between Iris, Horus, and we really became obvious. Before that time we had been the best of friends." Mom sat down at the table and reached for an open space on the table. A fresh cup of coffee arrived as if by magic.

"Speaking of the friendship you and dad had with Seth's parents, you left a lot of information out. You slipped up and hinted that Seth and I were somehow arranged to be together, and that the birth of Skylar was a pre-planned event, and you still need to explain the comment on species. It really does relate to my daughter, doesn't it?" As I touched the table with my hand, an English muffin just manifested in my open palm.

"Getting the hang of this thing, are you?" Mom laughed and sipped her coffee.

"You can't change the subject. I'm angry that our family, and I mean Seth, Skylar and I, have been forced out of our home and that Skylar is under the total control of the Arcturian Elders, as well as under house arrest by the

Alpha Centurions. I don't care how much fun it is to imagine the sunrise on Bell Rock out of a pretend window on this planet. I'm not impressed with the virtual kitchen and the 'Peter Pan' menu. I'm ready to gather up our daughter and vibrate to another galaxy. There must be someplace not controlled or touched by the Galactic Federation." I knew that as I ranted, my face became more and more animated and colorful. Mom sat quietly sipping her coffee and now enjoying a piece of New York Cheesecake.

"There isn't any other galaxy, my dear. We are not under house arrest; we are in protective custody. Skylar is getting the education you denied her on Earth. She's special. Let her come to full bloom before plucking her out of her new home. You can find out about the species thing yourself."

"This is not her home. I did not deny her an education. Mom, she was at a high school, almost at a college-grade level when Iris and Horus came to visit. You know that. Maybe you are as much a snake in the grass as Iris." I slammed the coffee cup on the table, and it vanished. I was so frustrated that I stormed out of the kitchen and out the front door to the peach-colored sky and abstract walking path of Alpha Centauri. I looked up at the plasma dome and noticed an orange sun rising above the crystalline city skyline and another very small yellow orb in the distance. I knew Alpha Centauri had three suns, but the sunrise on Chiron was beautiful. The sky was now a peach color, oddly tranquil. For a moment, my fear, anger, and frustration quelled. My mind clarified and focused.

I moved down towards the tram walkway with a new purpose, a direction for my movement. I spoke to an odd-looking woman with translucent octagonal scales on her skin. She had almond-shaped eyes of pure gold that were cat-like, slits up and down, not round. She was tall and beautiful with fine slender features. She wore a long headscarf that flowed from the top of her elongated head to midway down her long spindly legs. Her caftan was of a soft blue wispy fabric, like spider webs formed into clothing. I did not mean to stare but couldn't help myself. "Where can I find the Archival Data Center?"

The alien woman looked at me with the same odd expression I think I had when I stared at her. The woman looked down on me, and I watched her cat eyes scan my jeans, t-shirt, and tennis shoes, and then my face. She looked down on me as though I were a small child.

"Dear Earth child, you will need to have a guide and a permission slip to access any data. You should go home and ask the authority in your domicile." She turned and stepped onto the tram.

I was so upset that I don't even remember walking back to the apartment. I stormed in the door, "Mother, I need to speak to you at this very moment."

"Anna, I already know what you are mad about. You should have waited for me this morning. We do not have permission to just access archives without the proper authority."

"And who is that proper authority?" I demanded in a rather loud and annoyed tone.

"Don't get mad. Sit down for a minute. It's Seth."

"Seth? What the hell, Mom. Why would he have to give ME permission? It's my genealogical archives I want to access. I really do need to know everything, including the DNA of everyone in this family." I couldn't get that statement, species thing, out of my mind.

About that time, Seth and my father came into the kitchen and sat down. I imagine they heard the whole conversation. I wasn't exactly quiet.

"Anna, I will take you to the archives, but let's at least allow your parents the opportunity to fill in the blanks on your DNA."

"What makes you think they will be forthcoming? They haven't bothered to do so to date, Seth. I can't tell you how absolutely pissed that makes me." And with that,

I gave my mother a look that should have knocked her to the floor. She did look to the floor and fell silent, like that was a surprise.

"Anna, let your mother have a moment to explain. You will have to open up your mind and really expand your Earth girl brain. Seriously, the information will be very hard for you to wrap your mind around." Dad rubbed the back of my mother's neck and looked apologetic. "I suggest you get a pen and paper for this." My father, always attempting to be the sound of reason.

"Right, they just keep that kind of antique writing tools around in this galaxy." As I spoke the words, a tablet of paper and a pen arrived in front of me on the table, George Jetson style. I picked up the pen and glared at my mother.

"Okay, well, um, let's start with me then. Hello, my name is Diana, known in ancient Greece as the Goddess Diana, the huntress. Wipe that stupid look off your face and listen! I am a Titan. My parents were Zeus and Leto, although Zeus was mated to Hera. My mother, Leto, really made Hera mad. My mother had to hide me and my twin brother Apollon for fear Hera would kill us. Mother sent us to live in hiding with sea nymphs. We were there so long that Apollon mated with a beautiful mermaid named Hekate and your aunt Skylla was born."

"Off topic, Mom. I already know about Skylla, sort of, so go on. I'm all ears."

"So now it's your father's turn."

"Right, were you related to Wonder Woman too, your sister maybe? Can I have a real grasp on you guys' family history? Really. Is that all you are going to say, Mom? Not a real DNA chronicle. It sounds a lot like you're writing a comic book, not telling me the truth. Could the truth be so hard for you to tell me?"

"Time, dear, that wonderful ability to move within the folds of time. Just give us your open mind and listen." Mom reached across the table and touched me on my forehead as though she could just plant the information there and not speak of it.

"So, while we are dragging out our puffed-up resumes, I was originally called Hermes, son of Zeus and Maia. Same father, different mother, raised in a different world, Heliopolis. During my earliest memories, I was a messenger for the Olympian Council. Then I was off to the Science Academy on Titan. Yes, there really was a planet where the asteroid belt is now. That is where I met and fell in love with your mother. After a near total destruction of the planet began with some of the scientists who were in league with the evil six, it became clear that Titan was destined to implode. Some of us from the

Academy chose to return to Earth during different time eras. I'm getting ahead of myself here… sorry. Anyway, I was called Hermes, son of Zeus and Maia. Diana, daughter of Zeus and Leto."

"Does that make you and Mom brother and sister?"

"Not really, half maybe. Look, back then, it seemed as though there was a rush to populate the planet with our genes. It seemed to be some kind of hormonal drive, particularly Zeus, who possessed an enormous ego and raging hormones. He must have had hundreds of mates and consorts and thousands of children. But this is off track; let's get back to the genealogy of you."

"Where did you get the name Paulus from anyway? Just make it up or something?"

"No. During a seminar on spirituality and Earth-based religions, I became interested in the son of God mythology. I traveled back to the time of Jesus and became Paulus, or Paul of Tarsus, and traveled with Jesus. I was discovered while performing what the people of Lystra considered an act of the Gods. They wrote that[6] 'the gods have come down to us in human form.' They really didn't know how true that was. It wasn't too long after that that the destiny written for Jesus was fulfilled, and I returned to

[6] Acts 14:8-13

the Science Academy. I sort of liked that name and kept it."

"Not only am I a descendant of mythological Greek Gods, you add insult to injury by dangling biblical references at me. You hinted at a royal bloodline — mine or Seth's?" Frustration and a total lack of understanding must have washed over my face by now. An Earth girl suddenly related to the Greek Gods? Give me a break! And then to add Apostle Paul into the blend? My brain was yelling, bullshit!

"Anna, give your father a chance to map this thing out for you. I heard your mind, young lady."

"I am not a young lady. I am a mother, and I am mad." Seth suddenly moved his chair next to mine and wrapped his arm around my shoulder and his hand on my thigh. I was a little worried.

"The answer to the bloodlines is simple. We really aren't royal per se, but higher up on the command post."

"So, let's just jump me back in time to watch my father pretend to be an apostle."

"You are in the direct lineage of Uranus and Gaia, the mythical gods, the father and mother of humanoid life, at least within the context of Earth history. I am in the direct line of Ra and Neith, the highest office in the

Council of Worlds hierarchy, part humanoid and part reptilian. But both of us are hybrid blends of so much more. Our daughter Skylar is a blend of our bloodlines with added blends of DNA from all over the cosmos, becoming the royal child — The Universal Child."

"So, are you a reptilian? I thought you were from Tau Ceti."

Jeff stopped. It was far too much information on his genetic bloodlines to really grasp what that meant for him. He grew up in Arizona. His father was a learned professor of history, and his mother was an avid archaeologist. This is where he got all of his drive to discover the history of the past. All the other information seemed to be really unbelievable. So, who is his grandmother, and how does she fit into this whole time warp thing? Jeff couldn't stop reading, although he really needed to get ready for his meeting.

"Anna, I think I should take you to the Archives. It has been a long time since I researched the family tree. My mother was a purebred Reptilian, and they make lousy mothers."

My mother jumped into the conversation for a moment with, "That is the species thing I was talking about." Seth held up his open palm to Mom, and she looked back down to the floor silently. Weird.

"I spent the majority of my early years with my grandmother, Miramot on Tau Ceti. Sorry, Diana, I didn't mean to stop you, but not all Reptilians are the same. Those from the Pleiadian Council and some from Orion are just fine. Never mind, I know you are a visual learner, so a trip to the Data Archives will tell you everything you need to know. But in the morning, early if you don't mind."

Part of my mind was thinking about the time back in middle school when Seth was bullied and wondered if there was an innate tendency towards cultural and species bias even within the Federation. Were reptilians maligned universally, or was there something to the species thing? My mind wandered back in Earth history and the story of the snake in the Garden of Eden. It was curious: this division of humanoid and reptilian/dragon and Zeta (the evil six) which caused the Great War. I couldn't stop thinking about all the information. It seemed implausible, but my parents had been telling me this story my whole life, only they had framed it in a science fiction novel form. The novel was making the impossible seem probable, like Star Trek and Lost in Space.

I was living in a domicile on the planet Chiron in the Alpha Centauri constellation, a planet with three suns. I have a daughter that can float above the floor, fly, and telepathically tell you things, and an alien for a mate. What was the true reason this information sounded weird? My whole existence was weird. Why was the concept of marriage so foreign to the ancients and the aliens alike? Seth is more than just my mate. Seth is my soulmate, my life, my husband, if only in my heart, not on any archival document.

My mind rambled with a thousand disjointed thoughts. It made my head hurt. I wondered if the human mind was too small to venture out into the world of infinity, of time travel, and of a multitude of diverse species that occupied the ever-growing cosmos. The notion of science fiction, not science fact, now seemed more appealing. I closed my eyes and attempted to block out the information, to wake from a weird dream back to reality.

Skylar came in the side door without making a sound. She just materialized on the sofa next to me with a meow. I jumped like a scared rabbit, she laughed and, I have to admit, so did I. Skylar took my blue contemplative mood and brightened it right up. My daughter was real, so therefore so was my reality — not a dream, not a story, but a chapter in my life.

"How was school today?" I was sort of happy with the question. I had wanted to ask that question for a while, but the thought of her attending public school always struck me as irrational. Her gifts were too powerful and barely under control. Yet, here she was, on this alien planet, interacting with Earth children and alien children alike, and having a grand time with her peers, or at least with children her own age who could understand her.

"I had a great time in biology today. I created a new pet. Want to see?" she said.

I was almost afraid to say yes, but her sweet face melted my reservations and I nodded. Skylar reached into her pocket and pulled out something so small it fit in the palm of her closed fist. As I leaned in to look when she opened her hand, I let out a squeak and recoiled. Skylar laughed.

"It doesn't have any teeth, Mom. It won't bite. Come and hold it. You'll really love it."

I moved in and she pressed this warm furry thing into my hand. As I brought it closer to my face, it looked up at me. I saw large bright green eyes sticking out of a cream-colored fur ball. I couldn't find a mouth, ears, a tail, or legs. The weird little creature sort of resembled a tribble from Star Trek.

"I call it Rey. It isn't a girl or a boy, so that seems like a nice name," she said.

"Honey, I don't see a mouth, ears, or a tail. And, legs? What does it eat and how does it get around? Can it hear you?" I asked.

"You're silly, Mom. Rey feeds on love and positive energy, and it feels thoughts, so it doesn't need ears. It can float on vibrations, so it doesn't need legs. But it can sing and make pleasant sounds that make you feel happy. It's my gift to you. I'll make another one for myself tomorrow, maybe a blue one with purple eyes," she stated. With that, Skylar was gone, heading somewhere, possibly the kitchen, or somewhere… I wasn't sure. I looked at the silly creation in my hand and it blinked at me. A tiny birdlike purr emanated from it, and it floated off my hand and onto my shoulder, burrowing under my hair to my neck. The warm and soft creature gently vibrated, and my ears picked up a pleasant sound. It wasn't recognizable but beautiful, and I smiled. She was right; it made me feel better about everything.

The day was very beautiful. I was watching the sky out the virtual window, which reflected the weather back home in Sedona. The dome that protected the city outside kept the atmosphere at a constant 72-degree temperature. The virtual window reflected the winter scene of almost pink snow against the towering red Bell Rock.

It was nearing the holidays on Earth. The traditional dinner would be short of family members this year. Adam and Janus of Epsilon, currently living in Washington State by Lassen Peak, wouldn't be able to join us. That area was both an entrance to the volcanic tunnel highway system of inner Earth and also a portal. Despite living at the airport, so to speak, they would not be present. The only other family member not available was Iris, but I must admit, I hoped to never see her again. Since I discovered Seth lived on Tau Ceti with his grandmother, I didn't feel too bad about the monster-in-law being absent.

I was looking forward to my trip to the Data Archives Center in the morning, and that was about all I could think about at dinner. We sat at the table, listening to Skylar. She babbled on about learning to manipulate genetic material and create her own new species. As she continued to gush about her little creature, the furry bundle, still perched on my shoulder under my hair, let out a little giggling sound audible to everyone. Seth looked at me, and I parted my hair. Rey opened its eyes brightly and floated towards Seth. He extended his hand, and the little creature sat in the cup of his palm.

"What an interesting species, Skylar. What is its classification? Have you given it a name?" he asked.

"I named that one Rey, because it was for Mom. But the classification I put on my notes was Heartbradome. That's because I created it with cellular tissue of the human heart, the brain cells of an Arthurian, and the scalp cells of a mink. I am working to map out the genetic codices to change eye color and hair color. The eyes were that of a feline species, which are usually yellow, blue, or green, but I want purple, so maybe I should use grandfather's ocular cells. Just kidding, Gramps, they don't have them in the sample cell jars."

"Skylar, eat your dinner. We are all very proud of your accomplishments, but enough is enough," Diana stated as she held out her hand for the creature to visit her. As she did, the little creature floated back to my shoulder and nuzzled my neck. I felt an instant brightening of my emotions. I guess I missed the little creature after all, and I think it had bonded with me anyway.

"Grandma, would you like me to make you one as well?" Skylar asked.

"No, honey. Two in the house should be more than plenty. We don't want them to go hungry for love, do we?" Diana responded.

"That could never happen in this house, Grandma," Skylar assured her.

It was true; there was plenty of love to go around in this house. I was so happy to be together, my family unit whole again. We were off-world, in an alien galaxy, but we were home. Home is where the heart is, or so they say.

# CHAPTER 20

## The Book of Revelations

Seth had obtained the proper paperwork for me to join him in the archives. He had to visit the main security complex outside the dome to get clearance. I felt somewhat irritated, as if I were a second-class citizen. The document looked official enough, and I was happy to see the translation that accompanied the permission slip.

فرع" إلى الوصول هذا بموجب يرد أورانوس بيت أنا أن يعرف أنه

.آخر مكان ولا أرشيفية بيانات استرجاع مركز من "الوراثية

السنة من عشر الحادي الشهر من عشر الخامس اليوم هذا بيدي مؤرخة
.1977 عام الأرض

القوس مايكل الكابتن

*Be it known that Anna of the house of Uranus is hereby given access to the Genetic Branch of the Archival Data Retrieval Center and no other location. Dated by my hand this 15th day of the eleventh month in Earth year 1977.*

*Captain Michael Arch*

As we moved on the sidewalk tram, I held the paper in my hand and looked at Seth.

"What language is this? Is it Alpha Centauri?" I asked.

"No, it is Arabic, which is the language spoken by most of the Federation for some reason. Maybe because a great deal of the history of Earth occurred in the Middle East. Cuneiform is limited in words and is a dead language," Seth explained.

"Why would the history of Earth be of such concern to the cosmos? We've always believed ourselves to be the center of attention, but we're such a tiny speck of dust in the vastness of everything. Why then use an Earth language?"

"It wasn't originally an Earth language…it was a reptilian language, the first species to colonize that planet. So I guess that is why it remained the core language even though we've been involved in an ongoing war with the reptilian and draconian coalition for eons. Just because we appear to be so much more advanced than the people of Earth, we're still not totally civilized. Go figure."

We approached an ultra-modern looking structure of blue and purple-hued crystalline walls forming tetrahedron-shaped buildings. They were of the exact height and width but resembled the Giza Plateau against the pink sky. There didn't seem to be a main entrance, but the tramway stopped abruptly at the center building. We stood for a moment, and Seth looked at me.

"What?" I asked.

"Hold up the permission sheet," he said.

I was confused. "There's nothing to show it to. Where do you want me to point the thing, Seth?"

"Think window, woman. Point the front of the paper to the wall," he instructed.

I did as he instructed and, like the wall in the Lemurian tunnel, an opening appeared. We stepped in, and a tall, large-eyed, long-limbed creature with antennae stepped in front of me. It wore a long, flowing cape-like robe which looked to be made of a fine silver filament woven into a circular pattern. As my eyes floated upward from his robes to his angular long face, his eyes peered at me, and terror filled my heart. Rey, who was still living on my shoulder, gave what I could only interpret as a low growl or hiss. The creature stepped backward with a curious look on his face.

At that moment, I understood the power of this little creature created by my daughter. Rey was not only the core of love but my protector as well. Rey seemed to fend off anger, fear, or distrust - remarkable. I felt suddenly in control of my own space. A sense of well-being reappeared in my soul.

Seth stepped between me and the creature, giving him a quick look before announcing, "My mate and I are here to access the Genetic Data Archives. Will you usher us to our destination?"

The creature gestured for us to follow him with a sweep of his long arms. We followed him down a long hallway, which, like the ship we arrived on, had lights that appeared as we passed. I was struck with the thought that there were no pieces of artwork anywhere. I would have expected beautiful paintings gracing the walls of any museum or archival station on Earth, but here, there was nothing. It felt cold and uninviting.

The creature moved midway down the hall and stepped away, facing what was a solid wall of white milky quartz. Again, the wall opened to expose a doorway. Inside, there didn't appear to be anything but a bench and a countertop-style shelf facing a large screen.

"What was that creature, Seth? It looked like a praying mantis. And where are the documents we're here to look at?" I asked.

"That gentleman was an Elohim, one of the oldest and most learned species in the Cosmos. His species are still at war with the Orions in deep space. They're the ones responsible for the creation of the Greys, the slaves of the Annunaki. There are no documents within the Genome Data Center. Just sit down, and you'll understand," Seth replied.

I sat down next to Seth. Something pricked my arm as I set it on the table, and I let out a yelp. "What was that? Are there bugs in here?"

"No, it's an extractor for blood samples. It's painless. It just surprised you. It took a sample of my blood as well. Look up at the screen," he said.

I looked at the empty monitor on the shelf and then up to the main monitor wall. Suddenly, I noticed a document labeled "House of Uranus" appear on the wall. It was larger than life – my family tree. It wasn't a fancy-looking document but one written in Ancient Greek with an English translation included.

**ὁυσε οφ ϒϱανυς θε εξαλτεδ ϰομμανδεϱ οφ τιταν ολυμπιαν φοϱϰεσ**

**ϒϱανυς τὲ ἰγέστ ϰομμανδεϱ οφ Τιταν ανδ ωιφε Γαια**

**Γαια βοϱε α σον Κϱονυς ωὁ ωιθ ωιφε Ρέα ἀδ σον Ζευσ**

**Γαια βοϱε α σον Κϱονυς ωὁ ωιθ ωιφε Ρέα ἀδ σον Ζευσ**

**Ζευς ωιθ κονσορτ Λετο βορν σον Απολλο ανδ δαυγ-τερ Δαnaιανα**

**Δαυγ-τερ Διανα ανδ Παυλυς ορ έρμες οφ τέ όυσε οφ Ρα τογετέρ**

**Κρεατεδ α δαυγ-τερ Αννα σον Αδαμανδ**

**Αννα ματεδ ωιθ Σεθ οφ τέ όυσε οφ Ρα ανδ Ανυ ανδ βορε υνιυερσαλ χιλδ**

**Τίς χιλδ ωιλλ βε καλλεδ τέ όπε οφ τέ κοσμος ναμεδ Σκυλαρ όπε Αδαμυσ**

The House of **Uranus** was led by the exalted commander of the Titan Olympian Force. **Uranus** was the highest commander of **Titan** and had a wife named **Gaia.**

**Gaia** bore a son named **Cronus**. **Cronus**, along with his wife **Rhea**, had a son named **Zeus. Zeus**, with his consort **Leto**, had two children: a son named **Apollon**, and a daughter named **Diana**.

**Diana** and **Paulus**, or **Hermes** of the **House of Ra**, came together and created a daughter named **Anna** and a son named **Adam. Anna** mated with **Seth** of the **House of Ra** and **Anu**. From this union, they bore a child that held universal significance. This child would be called the hope of the cosmos and was named **Skylar HopeAdamus**.

The screen went blank for a moment and then a new image appeared.

House of Anu

**Anu, of the Reptilian house of Orion**

Anu joined with a woman and had a daughter, Nimsun, sister of Enki and Enlil. Nimsun then joined with the Serpent King, Ningiriddda-Zu, and had a daughter, Azimu. Azimu joined with a man from the House of Cain, Gestinanna or Noah, and had a son. This son, Japheth of the House of Cain, fathered Istar who joined with the bird man, Gilgamesh the traveler, and had the serpent king, Quetzalcoatl.

Quetzalcoatl, the Bird King, took a woman named Ninth and had a daughter, Kali. Kali joined with the Serpent King, Dragon King Aoshun of the house of Abzu, and had a daughter, Iris. Iris joined with the Sky King, Horus of the

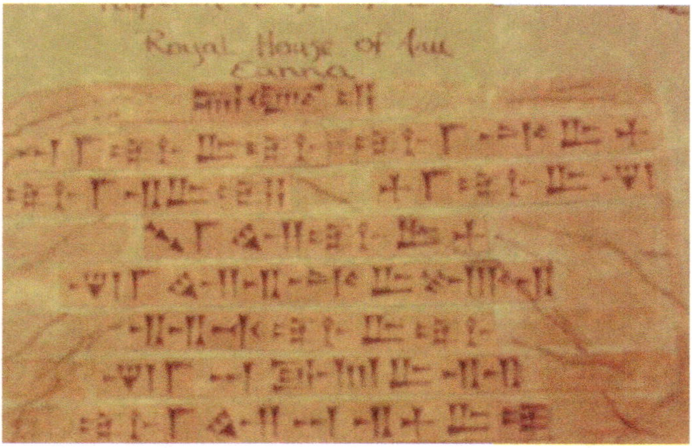

house of Uranus, and had a son named Seth. Seth joined with the daughter of the Sky King, Anna, the daughter of Diana of the House of Uranus, and they had a sky child, Skylar Hope Adamus.

"Well, there are two names I recognize... Noah... really, and Gilgamesh, as in the Epic of... So, this is your Royal family tree?" one might ask.

"Let me clarify this for you," I'd answer. "Be patient for once, won't you? Noah comes from Nimsun's brother's side of the family. Her son, Japheth, with his mate, gave birth to Istar, and Gilgamesh gave birth to Quetzalcoatl, the feathered Serpent God.

The word 'god' is used by ancient historians, but the Reptilian race included all manner of serpents and winged dragons, with and without feathers. They didn't understand that there were a multitude of creatures within the cosmos created by the One, but they saw creatures with advanced powers and assumed us to be gods."

"Well, my lord," they might say. "Related to everything reptilian as well as to the house of Cain, as in Cain and Abel, and Noah and the great flood... this is a bit hard for me to swallow, but go on... I'm fascinated."

"Quetzalcoatl mated with Ninth, a simple human woman, and she gave birth to Kali. Kali married well for a hybrid and mated with Aoshun the Dragon King. They had three children, one of which is Iris, my mother, your monster-in-law."

"Funny... very funny. Dragon Queen, I can understand and serpent, I get but Noah... not buying it," they might respond.

"Odd but true, my dear, and the rest of the story, I think you already know."

The screen went blank again. I began to tell Seth how worried I was about the constant repetition of the family tree with the Reptilian race. I had been told that they were enemies of the humanoid species as a whole. As I started to get up, Seth must have heard my thoughts.

"Not every reptilian or dragon is our enemy. Some of them are family. Sit back down; we are not finished."

"Very funny, you snake, you!" I would retort.

"Ssssss, now you have to endure my father's side of the family tree. Are you ready for the topper?"

The screen flashed on with vibrant colors, and I suddenly recognized Egyptian Hieroglyphics. Skylar would have been able to read them in a flash but, fortunately, like the cuneiform, each was accompanied by English translations. I had wished I was able to be in the classes with Skylar and the Lemurians inside the inner earth.

House of Ra, Anu and Cronus for Skylar Hope Adamus.

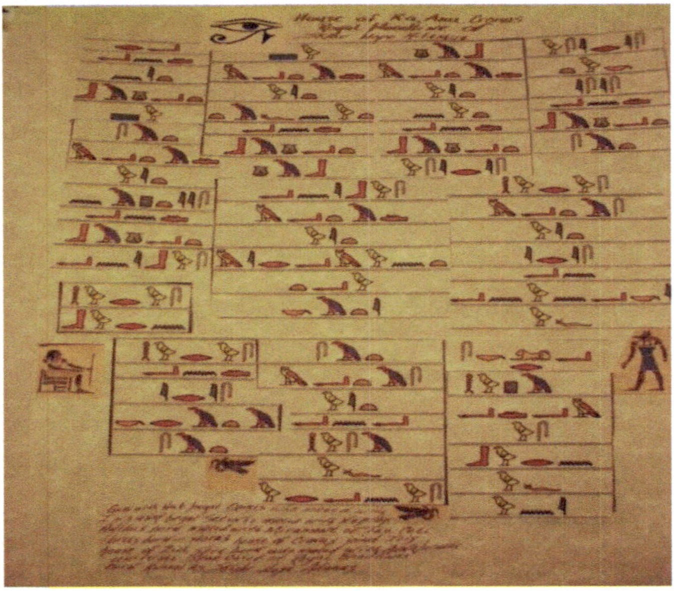

Ra and Neith begat Shu. Shu mated with Tefnut, begat Geb. Geb mated with Nut, begat Osiris. Osiris took Isis, begat Set. Set mated with Nepthys - Anubis was born, mated with Miramot of Tau Ceti. Horus was born. Horus and Iris created Seth who mated with Anna to create the universal child - Skylar Hope Adamus. Seth was part of the genome program at the Science Academy on Titan. Anna was also part of this program. Their mating was part of the prophecy of the Great Oracle of Delphi.

The meeting notes from the Council of Worlds, in compliance with the prophecy of the Oracle, stated: Hope was the product of royal bloodlines: House of Ra, House of

Uranus, House of Anu. She was blended with the life force of all, including but not limited to, the DNA of the House of Marduk, the House of Odin and the Draconian Royal bloodlines of the Abzu, a blending of both reptilian, dragon, and humanoid.

1. Skylar Hope Adamus, the Universal Child, represents the blood of those who have ruled and those who would rule from thereon.
2. It is her destiny to defend or to annihilate humanoid life within the cosmos, as written in her destiny.
3. She is the Universal Child, the bloodline of the Anuna (those of royal blood).

Signed by Helena, Ambassador of the Council of Worlds, Galactic Federation member, dated this eleventh month in the Earth year 1969.

The screen went blank, so did my mind. I looked at Seth, his expression was one of shock. Helena, or maybe Iris, seemed to have hidden agendas from our earliest meeting, our union, our daughter and the manipulation of our core families' DNA. Rey began to shiver on my shoulder and soon glided down to a pocket in my jeans. I wished I could have hidden away from this revelation, this genetic book of potential doom, an ending to humanoid life if Helena had anything to do with it. I laid my head on the shelf, Seth touched the back of my neck. It felt like the reality of the cosmos was suddenly in our laps, literally.

"Seth, what did that mean, line two about the destiny..." I asked.

"Not here Anna, there are eyes, ears..." Seth said softly and then in a loud and authoritative voice, "This is not your concern woman, we will return to our domicile now." Seth winked at me and I understood, at least I thought I did.

We made our way down the long hallway and out to the entrance. We stepped on the tram and aimed for our apartment. We stood silently on the tram, hands tightly wrapped together. There were so many unanswered questions, suspicion.

"Don't speak of this to our parents. This must remain our knowledge, do you understand?" Seth asked.

"No, I will not be sanctioned by you or the reptilian council here on this so-called sanctuary planet, not on Tau Ceti nor on Earth. This has become my quest for truth and I will discover all the information, with or without you, Seth. The fact that my mother Diana, from the house of Uranus, did not have authority to visit the Data Retrieval Center, but you, from the house of Ra, a known reptilian/draconian coalition, seem to have full authority everywhere. This tells me that this planet is not what it is reputed to be, it must be controlled by the Reptilian/Draconian Council and maybe you are either their pawn or a double agent. Don't lie to me, not one more time, Seth." As we reached the door I lifted my palm towards the main door and Seth suddenly pulled me away.

"Stop this, you are in danger, we are in danger, our family within these walls and particularly our daughter, are in dire danger. This isn't the time for you to just blurt out information or to rage against the structure of authority on this world. We are the guests of the Alpha Centaurians as well as the Arcturian, but this planet is like the United Nation on Earth, everyone here has an agenda. There is a hive mind at work here.

"One of the reasons you always thought that your gifts were shut off is because YOU have the power to turn all incoming communications off, it makes you hard to read, impossible to control and therefore very dangerous. In truth Anna, you are far more powerful than you know, you are just shut off."

"Shut off, who did this to me?" I asked.

"You did it to yourself the moment Skylar was taken, maybe even the moment my parents arrived in Arizona. You turn off every time strong emotions overtakes you. I think your subconscious mind knows what you are capable of. It isn't a bad thing; we will just have to hone your skills. Please wait with your quest, give Skylar time to achieve maximum power and then we will make our move."

"What maximum power, how long will that take before you think we can make our move? You are talking in some kind of code or something, I don't understand. What powers are you talking about and why can't I ask my own mother questions about my gifts, my family tree and the genome program we were involved in before our births?"

"Everything that occurs within the walls of our domicile is being recorded, discussed and voted on within the council chambers. I didn't want to make you feel uncomfortable but your first impressions were very correct, we are watched, through the virtual windows and beyond. We have to play the game and just wait for an opening to make our move. And yes, your parents are aware of the monitors, so stop being so mean to your mom." With that Seth placed his palm on the visual monitor device and the portal to our home opened.

Inside everything seemed to be moving on a normal path. Skylar was studying in the main room and my parents were in the sitting room relaxing in front of the virtual fireplace. Snow was softly falling in front of the picture window and the Red Rocks of Sedona began to become frosted with white and pink. My mother looked at me, she noticed I had questions in my eyes, she lifted one finger in her lap and waved no to me. I shuddered and stepped backward to the hall and down to our room.

I went to the wall, retrieved a robe, and entered the bathroom. This room lacked both a window and a mirror. I began to think, perhaps, it was free from prying eyes. Soon, Seth joined me and we sunk into the crystal bath.

As I started to speak, Seth placed his finger on my lips. "No eyes but plenty of ears," he said. "Relax, I have a plan to take us on an outing beyond the city's monitors."

Seth ran his hands down my body and we melted into one another. Time stood still. Passion flared and my heart swelled within my chest. Rey, who had become a constant extension of my soul, vibrated gently with our motion.

# CHAPTER 21

## Blow Out the Candles the Parties Over

"What is that thing?" The officer shouted as Rey floated directly in front of his face. He looked confused. His expression was a bizarre combination of terror blended with complete peace of mind. Watching his expressions change back and forth was amusing.

Skylar's birthday was nearing, and Seth planned a family outing. My brother Adam and his partner Janis received a coded memo instructing them to meet us on Janis' home world, Eridani, in the Epsilon constellation on November 20th. To get there, Seth had to exit the city and travel by an odd genetic bio-pod.

The Centaurians referred to this form of transport as a life pod. It was an individual plasma ball containing breathable, renewable fluid or mist. Each life pod, configured with the traveller's DNA, was also equipped with a bioluminescent outer field which cast light in the darkness of space. The pods were tuned to the traveller's mind and whisked them to their destination within a three-light-year radius from the original take-off point. This limitation prevented travelers from venturing too far. Eridani was 2.3 light years from Chiron.The plan seemed simple enough: travel to Eridani and set up a safe location unmonitored by the hive-minded council. First, however, Seth needed to stop at the security corridor outside the dome on Chiron and get permission to travel to Eridani.

On November 17th, Seth arrived at the security corridor. "Sir, I am Seth Adamus of the house of Ra," he began. "I am here to obtain permission for my family, including my mate's parents, my mate, and my daughter to join her brother and his mate Janis of the house of Zul. We are celebrating our child's eighth birthday and the bonding of Adam of the house of Uranus with Janis of the house of Zul. We intend to be off-world for no more than three rotations of the red sun."

The main officer of the corridor offered a guide for Seth, but he refused, claiming that his mate's parents were from the Academy and therefore versed in inner stellar travel. I doubted the truth of this statement, but I didn't care as long as we could leave the planet. In my heart, I suspected that the three rotations were a ploy and that the plans were to hijack a ship and escape, either to return to Earth or go on to Sirius, which, according to what I'd read, was an outpost or gate for a parallel universe.

Seth returned in three days, or so it felt by my time ratio. Skylar was so excited to see her uncle and meet the girl friend. She was so excited that she almost forgot that this trip would coincide with her eighth birthday. We gathered our possessions, or should I say a few odd clothing choices, and made our way to the main portal for transport. We had obtained a small shuttle to carry us to Eridani. Skylar had created her very own Heartbradome, a tiny creature even smaller than Rey. She named it Rain. True to her desire the little thing, Rain was a soft whitish blue creature with purple eyes.

At the main portal, an officer checked everyone for hidden contraband. I wasn't sure what they thought we could have managed to hide in a totally monitored environment, but then the officer ran his hands through my hair and Rey floated out with a hiss. The officer paced back and forth, looking as though he were fighting against possession of his soul. Before he could decide how to respond, Rain joined the scene, floating in front of the family like an irrevocable barrier of love.

"Look, I don't know what these creatures are, but they could be in jeopardy in the clean room. Who is responsible for these creatures?" he asked.

"Sir, I am the creator, and I assure you they will not be harmed by the clean room," Skylar responded. "They are biological entities and will ride on my and my mother's shoulders." She sounded so authoritative that I could not have been prouder.

"You... you are the Universal Child? Please forgive me. You, your creatures, and your family may proceed. I will monitor the health of your creations... if you will allow me." The officer bowed as if to royalty. I glanced at Seth with a questioning look. He returned my glance with a proud smile. It was then that I began to comprehend the significance of the royal bloodlines. It seemed that everyone here was aware of Skylar's lineage and viewed her as a princess with great influence over the people. I smiled back at Seth, acknowledging our shared pride.

Once again, we entered that suffocating, colorless room. With the whole family gathered in this small space and our Heartbradomes accompanying us, the fear seemed to disappear. "Close your eyes now and keep them closed," Seth reminded us. Suddenly, the room was filled with the sound of crackling electricity. Even Rey's fur puffed up until it was a round ball of prickly hair follicles. After the event was over and we brushed our hair down, the Heartbradomes still looked funny. They floated and rolled about, and I could have sworn they were making a sound like laughing. We were all laughing. They kept their eyes tightly closed, and as they rolled through the air, they resembled fluff balls dancing on a breath of wind. Skylar made a small, bell-like sound, perhaps in another language. It was vibrational, almost Lemurian in nature. The Heartbradomes opened their eyes and returned to the shoulders they seemed to call home.The door from the clean room opened into a silver-lined hall. We moved towards the end of the short hallway and into what appeared to be the main control area for the pilots. Seth ushered Skylar and the family into the seating area, and then he and my father returned to the main bridge. Seth powered up the ship and, as if all the light from the three suns was suddenly extinguished, we were in the darkness of space. I knew this was dark matter and that we would use it to propel us through the space-time continuum. We were off to the Epsilon Constellation. I was worried that this ship was too small to transport the whole family across the cosmos and beyond but I imagined we would likely have another ship ready for our grand escape.

It seemed that only hours passed until the ship slowed and the outer portals opened to view the planet we were now orbiting. "Seth, where are we?" I asked.

"Anna, this is Eridani. We will land soon and meet your brother, Adam, and Janus at her family's compound. It looks rather bleak from here, but it really is a nice place."

"Are we leaving from here?"

"Leaving? We just arrived. We are going to have Skylar's birthday party and Adam's bonding party together. I thought you understood that. This is the place where, as a family, we can openly discuss the past, future, and prophecies without any monitors. We will be returning to Chiron in two solar cycles. If we don't, they will send patrols to collect us. This is not the time for our grand exit, Anna.

This is an outing, a party, a celebration. Give some patience and time to plan. That time is not now."

"Why not now? We have a ship and we are already off-world. Let's get Adam and Janus and get out of here, please."

"Mom, I am not done with school. I want to graduate, and I want to return to Chiron. I have friends there. I never really had good friends on Earth; they couldn't understand me, and you were too afraid to allow me to show them the real me. You don't have any faith in humanity, do you?" Skylar seemed insistent, almost whiny, but she was correct.

I didn't have faith in either the understanding of most earth humans or their compassion. Listening to the rhetoric of the mentors and teachers of Alpha Centauri, I didn't have much faith in the civilized emotions of the Federation either. Bigotry and bias seemed to be a disease universal in all species. Maybe that was to be Skylar's gift to the universe: tolerance and understanding. She could become the creator of creatures both great and small whose life force was fueled by nothing more than the vibrations of love.

The shuttle landed on a long dirt pad decorated with what looked a lot like the Nazca lines. It made me ponder the history of Earth all over again. I didn't really know anything about this constellation or the indigenous life forms that occupied it, but maybe these Eridians were somehow related to the ancient travelers of Earth.

We stopped, and Seth opened the portal to leave the ship. We exited out the center of the craft by a light ray and walked out from under the ship into the waiting arms of Janus. Adam, in his typical fashion, lifted one arm in a nonchalant welcome. He never changes.

Janus was a beautiful woman. She, like the rather tall woman on Alpha Centauri, had octagonal scales on her cheeks of soft azure blue. Her hair was short, spiked up like a rock star. Her whitish-silver hair sparkled like iridescent crystals, needle-sharp, kind of like fine fiber optic thread. Rainbows and prisms of light shone off her hair. Her eyes were deep amber. She was as tall as Adam who was over six feet tall. I often wondered why I was a short five foot five inch shrimp. She was dressed in khaki pants and a tan flared shirt; she wore a utility belt and work boots. She looked like an archaeologist. I liked her right away, although I knew her lineage was reptilian or draconian. She had a warm smile and made me understand what Seth had told me about not judging all reptiles the same.

"Anna, I have heard so much about you and your daughter Skylar. Seth, I have known for many millennia. He and I went to school together way back when. Come, we have a transport waiting to take you to my parent's property.

You will be safe there." Janus rolled her arm around my shoulder, and Rey burrowed out from under my hair and floated onto Janus' shoulder. She looked and laughed, her bright smile beamed as she looked at Skylar, "Nice creature. Is this your first?"

"No, well yes, but I made two of them. Want one? I could make you one when we get back home."

"No thank you, but great little pet. You will meet your new grandparents in just a few minutes. Are you excited?"

"So, are you and my uncle a couple? I am not sure about being excited. I will have to judge for myself after I meet them, if that is okay with you."

"That's my girl," Seth said as he reached over and pulled Skylar to him. I found her comment to be rather rude, but Seth seemed to appreciate her honesty. Maybe that was a reptilian way: just speak your mind without thinking of what it made the other person feel like. I lived with a hurid (human/reptilian) for over a decade. You'd think I had this figured out by now, but I was never very good at filtering my emotions or comments either. Way to go, girl, I thought in retrospect.

I caught up with Janus and Adam and hit my brother on his shoulder, "So, you actually talked about me to your girlfriend, did you?"

"No, she asked. I told. You are not the center of the universe, you know. You always thought you were, but it isn't true, really. Right, Mom?"

"Don't fight, kids. You are too old for that now. So, Janus, are your parents expecting us? Was this trip okay with them? I don't want this family to upset their regular routines. I know they are very busy monitoring the environment and atmosphere on this planet and the new young binary planet Epsilon Indi 2."

"I am surprised you know about their work. Do you know them?"

"I think so, if they went to the Science Academy then yes, I think I do."

"Scientists, you are all alike. We will be at the house in a few moments, so hold your questions until then, won't you?" Adam said as he drove the transport towards the compound. I was struck with the similarity to the high deserts of Peru. When we approached the compound, I noticed the building material was made of the earth, much like the Hopi dwellings in Arizona. They seemed to be decorated with pictographs like the Hopi, and I began to wonder if symbols that represented the ancient pictographs were originally part of the Annunaki language.

We stopped in front of the earthen dwelling, and the door or wall portal slid open. Two tall humanoid-looking women stood in the doorway. Janus didn't look anything like either woman; I wondered who her parents were. I was confused.

Janus jumped out of the transport and rushed to the women, putting her arms around both of them proudly. "I would like to introduce you to my parents, Dena and Kara," she said.

"Welcome to our home," the shorter of the two women spoke, holding out her arms in anticipation of Skylar's approach. "We use most of the compound as an outpost for scientists in the realm who are also working on the terraforming of the new planet. We do have accommodations for you all, though. Let me get a good look at the Universal Child. Come here, my dear."

Skylar hesitated for a moment, but Rain, her creature, floated out from under her hair and towards the woman. I guessed this meant she held no hostile emotions, as that was how Skylar created these little fur balls.

"Just call me Skylar, if you don't mind," Skylar said. "So you are of the flying dragon reptilian class from the Lyra constellation…am I correct? I am also assuming there are no males living on this compound with you."

"Skylar, that is not a proper question," I cautioned, feeling embarrassed again by her just blurting out statements which seemed extremely rude to me. Was it possible that her parents, for our comfort, had shape-shifted to appear as ordinary humans, while Janus kept her draconian scales? I would just have to wait.

The taller woman spoke. "Anna, we are not offended. You have surmised correctly; we have shape-shifted for your comfort but we will return to our original appearance inside if you don't mind. As for Skylar's comment, it is such a pleasure not to have to explain to humanoid children the physicality of two females procreating. Skylar has been taught well for her age. Anna, you may not be aware that we are able to change our gender to procreate whenever the need arises. Dena and I have been on this outpost for hundreds of years and we wanted a family. We were so delighted to have a daughter, which is rare. Normally only males are born from this type of copulation."

"Well, I do have a brother, although he is much younger than I am," Janus piped in. "Adam is my baby brother as well. Is your brother here with your parents?"

"No, he has gone to the new academy on Arcturia to study linguistics. He wants to become an ambassador to the Omniverse Cooperative later. Skylar, he is about your age," Janus said, patting Skylar on the back and giving her a wink. I think Skylar must have been sort of taken aback by the information but she did seem curious.

"I have never heard of the Omniverse Cooperative, only the Galactic Federation. What is that anyway?" I was really beginning to feel like I was the kid, not Skylar, and that everything had to be explained to me. I think I gave my mother a look of total humiliation followed by disgust.

"Anna, sorry honey…my fault. We live in the matter universe; there are actually four known parallel universes called the Omniverse, but that could be incorrect at this point as we know the cosmos is ever growing; who knows what is out there now," mom looked sort of embarrassed as well as she attempted to explain this to me.

"Have you ever visited the other universes? Maybe that is the place we could escape to and raise Skylar without the Council of Worlds looking for us?" I asked.

"No, the other universe next to our own is the anti-matter universe. Everything there…we think…is a parallel yet possibly a polar opposite of our universe. We don't really understand how that works but we do know that we have a physical universe which lives in the third dimension and the ethereal universe which is in the fifth dimension. You can never travel there unless your life force has ended…we think, but don't know for sure. It is believed that is where the ethereal body travels to. There is no death, life… both physical and ethereal, made of matter or anti-matter, has a place in the grand cosmos…everything is vibration, energy which cannot be destroyed, the One or as the scientists on Earth called the String Theory," mom explained.

"Sort of preachy this afternoon, Mom. Let's go inside and discuss the cosmos and have some lunch…okay?" Adam said as he pulled Janus next to him. I was sort of happy to see him with someone he obviously loved, as he was such a student. I wondered for years if he would just be stuck behind some wall of monitors in some lonely galaxy waving as ships sailed by with that unemotional, un-impressed wave of his.

As we stepped inside the compound, I was struck with the impression that this was more of a science station than a home. There were beakers and glass enclosures with small plant life growing. We moved past the tables of organic life forms and down a plain-looking earthen hall. As we descended down the hall, the air temperature seemed to cool just slightly. The hallway opened into a comfortable living space with soft, desert tones of rugs and floor pillows scattered about. Janus sat on one large pillow in the corner and Adam joined her. We all followed suit and sat on the floor. One of the women, who appeared to have very deep scales on her cheeks and long hair which resembled feathers, returned from some hidden opening with a tray of food. Janus looked up and said, "Thank you, Dena." Lunch was served.

After we had our nourishment, the conversation began. Adam announced that Janus and he were planning on starting a family in the near future but that they were still acquiring advanced degrees in astrophysics and laser technologies.

"The combination of the two fields could become the new normal for energy and time travel in the near future," Janus proudly announced, bypassing Adam's announcement about family planning. I laughed secretly inside, realizing that she was the controlling partner in the relationship. Adam gave me a look that told me he had the ability to read my emotions. That was a bummer but not really a surprise.

"Are the two of you going back to Earth and California soon?"

"We will be leaving the same time you do, Anna, but we are hoping to see you in the future. Back home, I mean."

"Home...wouldn't that be nice. So, Seth, when will we be able to escape this place and find sanctuary on Earth?"

"Anna, we still have to return to Chiron in two days. I know you want to leave here and travel anywhere away from the Galactic Federation and the Council, but we have to wait for Skylar." Seth touched my hand, and I understood. We needed to let our daughter learn, to grow, and to come into her own power, over her destiny. I just couldn't shake the thought from my mind. Skylar, our loving child, could heal or destroy humankind, and it was written that it was her divine destiny to do so.

We relaxed and visited with Adam and Janus, tentatively making plans to visit their home in Mount Saint Helen's middle Earth complex. I was so happy to find that Adam had finally found his soul mate. I was curious to find out if they, or should I say Janus, really intended to have a family. Suddenly, I also wondered if Seth and I would have another child in the near future. Things did seem hopeful in many ways, but there were still major concerns for Skylar. I did suddenly realize that ignorance was bliss when it came to my life. The trip to the archives had been too much of an eye-opener for me, and I still couldn't stop my concern for my daughter in this alien environment, sanctuary city or not.

The trip back to the security center of Chiron seemed much faster than the trip to Eridani. It was like a horse returning home to the barn, speed was present. We moseyed out and galloped home.

# CHAPTER 22

## Fly Like an Eagle

As we returned to Chiron, my mind began to sing the new song by the Steve Miller Band, "Fly Like an Eagle". The lyrics rolled through my mind: time keeps on slipping, slipping, slipping into the future. I wasn't pondering what it could mean for our future, but rather, how time could be managed and manipulated. A myriad of ideas flooded my mind on how we could use the folds of time and space to fly like an eagle into the future.

It was right after we arrived back at Chiron that I became sure, beyond a shadow of a doubt, that I was carrying a child. The first time I had visualized my pregnancy, it seemed that I instantly gave birth. This time there was no one to interfere with time. I pondered the timeline of what was probably conception to the birth of Skylar and estimated that three months had passed. My mother had commented on my weight gain and I hadn't given it any thought then. Now, I was consumed with curiosity. Would this pregnancy be swift or would I endure nine months here in a controlled environment, the so-called sanctuary planet?

My desire to pack up our family and fly away was almost palpable. Now that I knew I was carrying another child, fear filled my heart – fear of the Council, fear of my mother-in-law, and fear for the future of this new child. I considered the possibility that this child could also be a target for the Council.

As soon as I had verified my pregnancy with the genetic clinic the following week, I secretly called my mother into my room to tell her the news. I hadn't thought to tell Seth first. Why, I wondered? Was my fear of what happened last time still so deeply embedded in my subconscious mind? My mother was my only female support system, but she wasn't knowledgeable about the gestation of hybrid children. Seth was a blend of Pleiadian and Reptilian/Draconian bloodlines, and I was mostly Lyran with genetically blended DNA. My mother and I both wondered how long it would be until the birth.

That evening, I planned for Seth and I to take a short plasma sphere trip to the outer satellite to watch the setting and rising of the red and orange suns. Reports from the observation platform suggested that this could be a very romantic and breathtakingly beautiful image for writers, artists, and lovers alike. We arrived just as the suns began their rotations. The sky erupted with the rise and fall of light, creating a burst of aurora rays shooting into the distant sky. The aurora between the two celestial bodies created a halo around the distant blue sun. I leaned in and kissed Seth, whispering in his ear, "You and I are expecting an offspring."

Seth wrapped his arms around me and pulled me so close to him that I felt we were about to jump time and space. "Anna, I am so pleased. Female or male?"

"I stopped at the genetic center and had a scan. I was told I would have a single child with two hearts or twins. I told them I did not want to know the gender because I wanted to surprise myself, have a normal pregnancy, a normal birth. Does it really matter to you?"

"No, not really. This time the child will be pure blood, just mine and yours. That will be satisfying enough for me, but we should see the geneticist at the clinic to make sure things are proceeding on schedule. Speaking of that, when do you think you will deliver?"

"Mom and I think maybe in three months or so."

"That's coming up very soon. We'd better make some plans, and quickly."

The family gathered as we announced the news of the upcoming birth. The geneticist at the clinic informed Seth and me that we would have a healthy offspring. We asked not to be told of the gender, but I was curious about the species. Two hearts could be aquatic, draconian or reptilian, but I kept my worry to myself.

Janus and Adam joined us and I mentally programmed the interior design chip to add a decorated Christmas tree to my virtual living room window. Skylar was so excited to see the tree that she almost forgot where we were. The holographic Christmas celebration and the announcement was short and to the point as Janus and Adam were back on Earth and resuming their studies. I decided to keep the tree up for a few more days just to make me feel a little more normal.

The New Year came and went without any fanfare. I was already beginning to show. Mother decided that the gestation was probably going to be between four or five months. The Genetic Clinician calculated that the birth should be in what would be early April on Earth. A few months, as the geneticist calculated, was not really enough time for Skylar to come to full power. I had hoped we could leave before he or she was born so that the child would be a native Earthling. Skylar was not born of Earth but was conceived there, which made her birth certificate seem okay.

But this child was neither conceived nor would it be born on Earth, making this child an alien, human or not.It was odd to ponder the humanity of my children when we would have been the aliens on the planet anyway – at least from what I learned in Lumeria and in the documents of the great libraries of Alpha Centauri. Earth beings were referred to as "primates", the indigenous inhabitants of Earth (Terra) were the reptilian and then the primates, or the Adama, the blended creatures. The rest of the so-called humanoids were refugees of the Great War, the geneticist that created "man". Maybe it didn't matter that my child or children would have an Earth birth record. My parents obviously didn't.

I spent my time on Chiron reading and learning about the fate of the universe, the history of the Great War and the information that my parents had documented in what I used to understand as their novel, "The Clash of the Titans". Skylar, like her instant birth, seemed to be advancing in knowledge and power daily, almost exponentially. I doubled in size and became very uncomfortable moving about, even on the tram byways.

As if by some exact calculation, I woke on the morning of April 24, 1978, as the first contraction began. Seth sent for the genetic clinician to come to the house. By noon, fraternal twins were born. We named our son Noah and our daughter Athena. The official records for the houses of Ra and Uranus were Noah Set Adamus and Athena Rhea Adamus. They were both golden-haired infants with large, deep blue eyes. Even as newborns, the ringlets of golden curls twinkled in the morning light. They were so breathtakingly beautiful, I was stopped by a diverse group of species from around the vast galaxy who marveled at their aura.

Skylar was at full power by her ninth birthday, and the twins were starting to walk at seven months. Seth and I took another trip to the observation platform for our annual anniversary. It was a place where we could speak freely, yet softly, to formulate our plans. The suns rose and fell with beauty as Seth explained the plans. We would travel to Eridani, and Dena would have a science exploration ship standing by. We would use this transport to go to Sirius and use the time portal. I wondered when we would arrive on Earth…and where.

The next day, Seth contacted the elders with the request to travel to Eridani for a family gathering and the initiation of the twins into the family tree. This gathering was a sacred ritual dating back generations upon generations and would not be viewed as anything but innocent. Family genealogy was paramount in the galaxy and bloodlines the absolute ranking of power families. We were part of the royal bloodline, and this initiation was not only okay but anticipated by the elders and the council.

I was to pack, but not appear to take much. We couldn't appear to be escaping the sanctuary after all. I used Skylar's old toy chest to pack our few important belongings. There was a hidden floor in the chest which held our documents and important family information and my leather journal. I placed Skylar's favorite book inside on top of the clothes and possessions for the twins. Skylar packed a small bag with her genetic code book which she wrapped in a pair of jeans and a t-shirt. She remarked how vintage she looked; it made me smile. She looked just like I did not too many years ago. Seth and I took a small bag with just enough clothes to last for three days…the amount of time Seth requested for the round trip.

Two days later, we arrived at the security center where the guard looked at the toy chest with the pink bunny on it and appeared confused. Seth opened the chest for the guard, simply stating, "Earth children…can't leave home without their toys."

The guard gently moved a few items on the top of the chest. "That was my favorite book, and I read it to the twins every night. Isn't it a lovely book?" Skylar interceded.

"My little girl likes to be read to as well…odd you would think she would like to read the monitor but maybe it is the personal touches that are the most important…don't you think?" The guard closed the toy chest and winked at Skylar. She smiled brightly at him, and he continued to inspect the rest of the bags. "Well, all clear…enjoy your initiate process with your two beautiful children. May the One be with you."

He hadn't noticed or possibly decided not to confront us about Skylar's creations, her Heartbradomes which now numbered four. His demeanor indicated he knew that Skylar was the Universal Child…maybe that alone gave us the pass. It wasn't the tear-everything-out-of-the-cases we had seen as we approached the security area. I was happy that the fear of the Universal Child helped us pass without notice.

We arrived on Eridani, and Dena had everything ready for our departure. We had just enough time to introduce the twins to her as we boarded the science vessel. We were off to deep space, using the inertia of Dark Matter to propel us across light years. Dena had provided the children with enough changes of clothing to make it the nine days in space. Dena requested Skylar leave her Heartbradomes with her as she felt it unsafe to travel with them. Skylar was sad but understood. She even provided us with uniforms that appeared to be replicas of the Titan Space Academy. According to Seth, Sirius had a time and space portal unlike anything in the known universe. It will be with this portal we decide when and where we arrive back on Earth.

When we approached Sirius, or Canis Major, and landed on the planet Mirzam, we were greeted by a tall, odd-looking humanoid creature with light blonde hair and reptilian eyes. He seemed to already know of our plight and was ready to usher us to our destination. I thought maybe Dena had forwarded the information, but Seth told me in my mind that he had telepathically reported our arrival and situation.

"Seth, I want to go home…I want to be near Sedona. Is that at all possible?" I was homesick for my old life. I was so happy there when it was just us…no destiny or plots against humanity…no Council and no control over our every moment.

"There is a portal within the Southern rim of the Grand Canyon. It hides the corridor to inner Earth. That is about as close as I think we should get…at least for a few years anyway."

# CHAPTER 23

## Let's Do the Time Warp Again

The theme song of "The Rocky Horror Picture Show" perfectly fit the situation we were about to embark on. Seth had chosen the year 1900 for our return to Earth and selected the entrance to Cibola as our destination. I trusted him with the details. We prepared our children for the warp and warned Skylar that she would feel the darkness and cold envelop her. It would seem to take ages, but in truth, only moments would pass. Seth held Noah, and I held Athena. Our possessions were condensed to fit within the toy chest. We entered the portal, sitting motionless on the toy chest, dreading the moments when the darkness would hit us. The void would be calling, "Let's do the time warp again.[7]"

The darkness and cold wrapped around us, and the air seemed to be sucked out of our lungs. Athena began to whimper, and I could feel her body tense. Instantly, the candlelight returned, and we were within an earthen room. I could hear chanting just outside the door. It sounded Native American. I must have looked puzzled.

---

[7] 1973 movie adaptation of play, Rocky Horror Picture Show. Music and lyrics by Rock Horror called Let's do the time warp again.

"You are hearing the Zuni chants to the sky gods," Seth explained. "We are in the home of the Anunnaki. They were responsible for the Zuni and Hopi tribes' survival of the Great Flood. Every year the elders make the treacherous climb into the cavern to pay homage to the gods. We must have accidentally picked the wrong day to arrive. Keep the children calm, and I will try to find one of the Anunnaki elders."

Seth left via a wall ladder to the ceiling. I wondered how he knew it was there, as I hadn't noticed it until he began climbing. Skylar looked up in awe. I was worried she would laugh. We sat very still, and soon the chanting stopped, and the sound of the drums and rattles that echoed through the walls came to an end. It seemed the Zuni had left the area. I handed the twins to Skylar and peered out the opening. A large meeting room lay in front of me. Billowing smoke from the center fire pit rose to the tall cavern ceiling as the wood continued to smolder, but the room was empty. Seth had not returned. The silence was suddenly deafening. Skylar entered the main room, and the twins began to run around the main fire pit as though they watched the Hopi and Zuni elders dancing in a circle. Perhaps they had envisioned the scene.

The smoldering wood was slowly dying out, as was the light within the cave. I was hoping Seth would return, as I was not happy with the destination. From somewhere above me, I heard something—a muffled sound of shuffling. We all went silent. From an opening in the far wall, opposite the room we had just been in, I could make out the outline of a man. His shape looked like Seth, followed by two other taller shapes. Then a beam of light illuminated the surrounding cave. Hieroglyphics in full color appeared on the walls. I thought we were going to the Grand Canyon, not Egypt. "Seth, is that you? Where the hell are we? I didn't want to be in Egypt."

"Anna, I'd like to introduce you to our hosts, the Anunnaki," Seth responded. "We are in the Desert Southwest, just as I told you. We are inside the Grand Canyon, on top of the ancient city of Cibola—or at least one of the cities called Cibola. We are in 1901, and we are safe. The room we were in will be our domicile for a while. They have all the furniture, food, and water we need. We're going to have to entertain ourselves, but beneath us is a library where we can continue Skylar's education and begin teaching the twins languages of the ancients."

"What's with the Egyptian hieroglyphs? Wrong area for that kind of language," I responded. Jeff laughed. That was his first thought too as he looked at the Egyptian art on the walls with ancient pictographs and hieroglyphics. It was a true anomaly.

"Walk around the corner, and you will see pictographs left by the early Zuni/Hopi people who lived here for a decade," Seth explained. "It's a repository of ancient languages and religions saved from the Great flood, which decimated the Earth. The flood was just one of the extinction events which were orchestrated. There have been five extinction events, and one almost completely wiped all life off the face of the earth. This is a remarkable library. Wait and see. It's a wonderful field study for Skylar."

"Not much of a home for the rest of us, though," I countered.

"It will be, trust me. I wouldn't let you rough it that much, your royal self," Seth bowed as if to a queen. He wore that silly smile on his face—the one that was supposed to show humor. I was not amused. We were stuck in the time warp again, and I was not a happy camper. I didn't know how to get out or how to survive in the past, so I trusted that Seth would be able to navigate the new time zone.

Before I could complete the tour of the pictographs and hieroglyphs on the walls and wander through rooms with the golden statues of Ra and Buddha sharing the same hall, I heard the sound of many feet moving quickly near the room we had arrived in. The children and I moved back through the corridors towards the main room and looked inside the cubicle that was to be our home. It was suddenly furnished with woven straw rugs and ornate gold-leaf settees. The room was divided by wall screens adorned with fine Sumerian artwork. The lighting was modern, almost like the crystal lights of Chiron. It was comfortable. Skylar had her own area, and the twins had a smaller section where the toy chest rested between two comfortable daybeds. We shared the main area, with the large settee serving as our bed and our daytime sitting area.

I walked back out to the main room. Instantly, tables were arranged in a circle around the fire pit left by the Zuni. I marveled at the speed and efficiency of the Anunnaki. A tall man moved towards the table. He placed trays of fresh tropical fruit and nuts on the table and indicated we should eat. Beakers of ale were set beside chalices.

I asked if they had plain water. He pointed to the beaker. I looked up at Seth and said, "Traditionally, beer is served to all ages. It is a safe, germ-free drink the Anunnaki, as well as all the ancient people, drink. It's not really alcoholic. I promise." Seth smiled.

"Well," I responded, "the kids will sleep soundly at any rate. It'll probably be good for me as well. Bottoms up!"

# CHAPTER 24

## Crystal City

I woke to the sound of my children laughing in the main room. Reaching next to me for Seth, I found he must have already gotten up. I had hoped he was monitoring the twins, as they were very little, and we were in a cave after all. I dressed in the jeans I had brought with me and a t-shirt and went to see what seemed so funny.

The main room had the tables set with beakers of what looked like fruit juice. There were dried fruits, nuts and, grain blended together with what looked like milk. Cereal for breakfast? How odd. The children were on the floor, playing with a small lizard. It reminded me of the tiny lizard Seth had me play with on my mountain in what seemed eons ago.

"Kids, come eat breakfast and leave that little thing alone. Where are Skylar and your father?" I asked.

"They went that way, mom," Noah said as he pointed to the west wall.

We sat down at the table. I didn't see any reason to wait on them as they could be a while. They were probably at the library. The white liquid I thought was milk turned out to be coconut milk. The cereal was very tasty. It was some kind of grain. The juice seemed to be prickly pear juice sweetened with honey. All in all, it was a very nice breakfast.

Before we had finished, Seth and Skylar came around the corner and joined us.

"So, what adventures were you on?" I asked.

"Skylar and I went to the west corridor into wing 1 to read the story of the Annunaki and Hopi legend of surviving the great flood. I thought I would wait until we could all go to the crystal city." With that, they both ate their breakfast and were ready to venture deeper into the cavern floor.

With Skylar holding Athena and me holding Noah, Seth led the family up the path of the west wing tunnel number one. We walked past the pictographs of the Hopi story. Then there were statues against the wall of the Naga, the blue "gods" of India, the Buddha, and the statue of Ra.

"Why are there so many effigies of religious figures from all around the world?" I asked.

"Anna, this was like a storage unit to protect the treasures and history of the world prior to the Great Flood. It was what the Council of Worlds thought worth saving for the future education of humanity," Seth explained. He continued on at a faster speed, making us have to practically run to catch up.

We finally came to what looked like the end of the line. Seth reached up and pulled a stone from the wall, and a door slid open. There was a waterfall flowing in front of a crystal step pyramid. It was rather dark, but Seth pressed his palm against the side of the wall to the pyramid steps, and light from the quartz crystals in the walls of the cavern, as well as the stalactites, filled the whole cavern with soft light. I could see beyond the pyramid to other crystal structures. It was a city underground. I wondered if this was the home of the Annunaki.

"Welcome to Cibola. One of seven cities that were built by the ancestors. They are from Giza to Peru and beyond," Seth announced.

"Is this where our hosts live?" I asked.

"Well, yes and no. They are stationed within the crystal city for now but as soon as we are ready to leave, they will go back to their original homes. This is more than just the repository for history and knowledge but also part of the energy plants within the inner earth. It is the reason we have a magnetosphere which is protecting the earth from solar radiation. These stations are all around the world and have been here since before the many ages of occupation here on this planet," Seth explained.

Seth touched a small oval indentation in the pyramid and the door to the base of the structure opened.

The family entered into the pyramid and soft lights appeared to come on as if motion detectors activated them. It was magnificent. Suddenly, an Annunaki elder appeared as if by magic.

"Welcome to the library. I understand that you, Skylar, are here to learn all that is within this single library. I will be here to be your interpreter and tutor," the elder said.

With that said, a crystal skull floated down from the upper shelves at the top of the pyramid and gently glided into the teacher's hand. He held it with great reverence as he handed it to Skylar. She instinctively sat down and placed it in her lap, looking awestruck.

"What am I to do with this?" she asked.

"Place one hand on each side of the temple of the skull and close your eyes. I am assuming you have been trained how to meditate and clear your mind. Once you are in a trance-like state, the information will simply fill your mind," the tall Annunaki explained, bending down and placing his hand on top of her head as if to start the process.

As she sat on the crystal floor, the rest of the family took a walking tour of the interior. Seth and I brought the twins deeper into the city of Crystal and they were so delighted with the prisms and the dancing of the lights coming in from the top of the cave. The streets were paved with crystal and mica and literally twinkled like diamonds. I was so awe-struck by the beauty of this hidden gem within the Grand Canyon.

Seth thought we should just leave Skylar to her education and go back home. It wasn't really what I considered home, but it was what we had to deal with while we hid from Helena. Even though the twins were barely eight months old, they, just like Skylar, were ahead in development for their ages. Athena was fascinated with the art and the pictographs within the different wings away from the center room. Seth walked with her, explaining the meanings of each area of interest. It was our way of homeschooling them while Skylar continued to learn all that was knowable. Noah was fascinated with the written languages on the walls. I taught him what I had learned, but it was really Skylar who became his tutor.

We arrived at the Grand Canyon in 1978, our time, but in this timeline, it was 1900. The only time I could smell the outside air was when I ventured to the main opening of the canyon and looked out onto the mesa, then down at the river below. It was the same sky as home, my Sedona. The climate was identical, the air crisper and cleaner, but still the sky of home. I would go just to view the sunsets and imagine being in our home, overlooking the red rocks of Sedona, while the sky transformed from pinks to purples and finally to black.

The children continued their education while Skylar advanced to almost god-like knowledge. It was almost frightening, considering her destiny. The Chiron library claimed she was to usher in the new age of enlightenment for the planet of children, or what I know as Earth. Then the other components of her destiny were to defend or to annihilate humanoid life within the cosmos. At the time, I thought that was the hope of the Orion factors, the reptilian council's wishes. Whichever the destiny, Skylar's future was frightening for me to imagine.

Every day rolled on in a regular pattern. It was so predictable that I was bored. Even keeping up with this diary became hum-drum. I was ready to use my pendant to go anywhere but here. I knew this was a silly idea, and we were hiding Skylar until she was ready to get on with whatever magical thing she was supposed to do that would change the world and the entire universe.

Seth and I were occupying the same small space as our children, so intimacy was a challenge. We had, on occasion, waited until they were sleeping and headed for the cavern entrance (my favorite place). There we would roll out a blanket, watch the stars, roll into each other's arms, and let our passions flow. That may also be the reason this was my very favorite place.

I became the math and science teacher for the twins. They excelled past my ability to teach them, although they were now only five years old. Time seemed endless here in this cavern of religious relics. Seth took on the job of teaching astronomy and ancient language arts. Athena's passion for learning leaned more towards the artifacts, the religions represented, and the history of their cultures.

At one point in our longest stay anywhere as a couple, one of the Annunaki came to add to the education for Athena and Noah. He spent a full year with them. Noah learned the art of cuneiform, the ancient Sumerian languages, ancient Arabic, Mayan, Aztec, and Lumarian, which was tonal, bell-like sounds. I think Noah liked the fact that he could irritate his sister with that language at night while she was trying to sleep. He always seemed to know when Seth and I snuck off at night.

One afternoon, while Skylar was on a mental break or perhaps she had learned it all, she got Noah to sit on the floor and scratch their names near the floor of their names. They were in cuneiform and Arabic, as well as good old English. The indents were tiny and not noticeable, or Seth would have had to remove them as to not leave any mark in an ancient dwelling. I was impressed that she had time to spend with her baby brother.

It was early in the morning when I heard a lot of movement in the main room. I woke to find the Annunaki removing furniture from the main room. She spoke in a hushed tone and said, "Pack up right now." Seth came out, as did the children, and we were all alarmed to see the activity in our home. We had been here for years. What was going on? Soon, they took all the furnishings from the living quarters away, and I heard what sounded like a blast. Seth had gone to the crystal city to make sure that nothing of ours was there and returned by way of the flap in the roof of our living quarters, climbing down what looked like invisible steps. There we were, sitting on the floor of what used to be our home. Now what were we to do? Had Helena and the Federation finally found us? Fear filled my heart.

I had my pencil in hand as I attempted to document what was going on when the Annunaki woman stepped in and said, "We only have about 10 minutes to make our escape." To where? Noah picked up the toy chest, and Seth grabbed it from him.

"They are coming, and we have to go. That will not fit through the hatch and will have to remain." He grabbed my diary.

Jeff flipped the pages, and that was the last entry. Who was coming, and what was happening? What happened to the family, and what happened to his grandmother's future and his history? His mind flashed back on the room that connected to the roof hatch. It was true that the box would not have fit through the opening, which he surmised is why it was found covered with a century of dirt. He wondered about the pictograph of the family pointing to the floor where the hatch was found while searching for a way out of the crystal city. Had his grandmother left that for him? Did she have future vision of what would be happening when he excavated the cavern? If she knew that much about future insights, did she know if he would be able to continue his work after the Regents' investigation?

He would have to wait until after his meeting to determine his next steps. He could hardly wait to be done so he could return to his grandmother's home to hear the answers to his copious questions.

# CHAPTER 25

## Let's make a deal

Jeff woke up at the crack of dawn to find his grandmother already sitting at the kitchen table. "Didn't that woman ever sleep?" he wondered. She had already put hot coffee, biscuits, and jam on the table and was waiting for him.

"I know you're meeting with the Regents," she began, "You will need them to carbon-date the so-called graffiti that has Skylar and Noah's names on the wall. It should clear up the suspicion that any of your students created it. I don't think they will allow you to continue your research this year, but you should be able to get on for next year into the other Cibola in Caneloria Caves outside of Guatemala City. Yes, before you ask, I do have some future knowledge. I also have knowledge of the former reality as well as the current one."

"So, if you knew what was going to happen, couldn't you have warned me not to let June come with the rest of her classmates? She could have been spared."

"No," she answered, "sometimes destinies just have to play out. It isn't something you can manipulate, as you have already guessed."

"Speaking of that, I need to know what happened. I couldn't sleep last night and completed reading your journal up to the point where you said what's going on. You were still in the Grand Canyon. So what happened?"

"Too much to tell you. I will say what you already know. It was 1906 when the cavern was breached by the Smithsonian. It is that breach which surprised the Anunnaki and forced us to leave so abruptly. Seth took my journal and placed it in the toy box. He had advance knowledge that you would find it. It isn't something that was supposed to be discovered, as what transpired after that is a forbidden knowledge situation. He just wanted you to have the full knowledge of your family's traditions and past. I didn't know he wouldn't be here to teach you all of this." With that last statement, a tear formed in the corner of her eye and left her silent.

Jeff wrapped his arms around her and kissed her on her forehead. "I will be back tonight. I hope you will tell me the rest of the story." With that, he gathered his briefcase and notes for the meeting and took a lift to downtown Phoenix.

His meeting was to be held at 10 o'clock. He rushed into the building, and the same young lady looked up at him and pointed to the lift. He knew where to go. The Regents were waiting for him. They all sat around an oval conference table looking very stern. One of them motioned him to sit at the head of the table on the far side by the windowless wall. "Nothing like making him feel trapped," he thought to himself. He pulled out his notes and waited for someone to speak. The silence seemed to be endless when finally one of the Regents stood and said, "Professor, you are here to explain the inconsistent record keeping and monitoring of your prized students.

We are most concerned about the defacing of antiquities recorded by your student June. We are sorry to hear of her passing."

Jeff couldn't help but think they had been instrumental in the events which caused the accident in the first place. Their way of cleaning up the evidence of history being leaked.

"Gentlemen, I think this information was already discussed at our last meeting. The so called graffiti has been dated, so that issue has been asked and answered. I thought we were meeting to discuss the continuation of the Grand Canyon research. This expedition represents extra credit and goes a long way with my students for their resumes. Does this mean we will be able to continue our research with my students sometime next week?"

"No, your students should be receiving good marks on their individual record keeping and should be able to be placed within the research and exploration units after graduation. I hope that works for you, Professor."

"I will let my students know, but I have my junior class coming up this year and would very much like to make a deal with you to continue research in my theory of the Egyptian connection to the lost cities of Cibola. I believe I have researched enough for you to invest in an expedition to Guatemala where there is a rumor of a grand city in South America inside the Caneloria Caves outside of Guatemala City. It would be a grand adventure for your future associates."

There was soft mumbling inside the conference room when finally, the head Regent stood and simply said, "Professor Ryan, we have always had great results from all your former students that are currently in the field. We think that your next expedition could be arranged. We will get back to you at the start of the new quarter at ASU. Thank you for your candor and your appropriate response to our inquiry. Good day." With that, they left the room to a side exit Jeff had not noticed when he entered the room.

He sat alone gathering his papers and out the door he went, down the lift. He straightened his shirt and smiled at the young woman at the desk as he left the building. She called after him, "Have a great day."

# CHAPTER 26

## This Is the End of the World as We Know It

Jeff returned to his grandmother's house after an almost successful meeting with the Regents. His disappointment was palpable, knowing he wouldn't be able to continue the exploration of the Grand Canyon cavern and, particularly, the crystal city. His grandmother had been correct in her insight. Yet, it was encouraging that the Regents were entertaining the idea of exploring Guatemala. That could be a real challenge, as the site had not been found to date, and Jeff figured his grandmother had insider knowledge of its location.

In the afternoon, Jeff returned to Sedona. The summer weather was heating up, and he was happy to escape the stifling heat of the Phoenix valley. Entering Gram's house, he found her sitting in the kitchen with a tall glass of tea for herself and another for him. It always surprised him how she seemed to know when he was coming or had arrived. She just knew.

Gram had her journal open to the last page. Jeff figured she'd read up to that point to refresh her memory. Despite her age, her mind was remarkable. "So, Jeff," she began, "I see that you are ready to hear the rest of the story of how the world ended and was recreated in your version of history."

"You'd better make yourself comfortable. How about we retire to the patio? I know it's a little warm, but there's a lovely breeze." Jeff picked up his tea, and they moved out to the patio. It was always comfortable out there. "I read the last page and so I think you would like me to fill in the rest of the story."

"I first want to know how much of your journal was fact and how much of it was fiction. There are so many things that align with science, but not in the correct era, and the amount of violence is unnerving."

"Well," Gram started, "I have to tell you that the majority of the journal was fact-based, and I lived it. I may have embellished some of the early impressions of my life as I was just a kid, but the amount of war and violence was real. The universal struggles and conflict in the cosmos, even though the record keepers don't want that information known, are real and ongoing. They believe society is better off, more placated, when the negative information is hidden from them."

"So, there you were, a young mother in the cavern of the Grand Canyon in 1906. You had to leave the box, book, and journal behind. Why didn't the Smithsonian find it? How did you get out of there, and where did you go?"

"Let's start at the beginning. The Annunaki, who had provided all the furnishings for our domicile, had to remove everything. Seth had informed them that the box had to remain, so they simply buried it in soil. Your grandfather had more knowledge of the future, your future, than I did. I didn't learn about everything that was to occur for years. You'd think I'd be in the loop of information, but like the time anomaly, my knowledge of the future and particularly my children's and grandchildren's destinies might have led me to try to alter the outcomes. I have to admit; I would have tried to stop your wife from going to work on the day she died if I'd had advanced knowledge."

"I wish you would have," Jeff said with a tear in his eyes. The grief and loneliness he felt were still very much present.

"Seth climbed up the wall ladder first, and while I was getting the children ready, he scratched the image of the family at the top of the tunnel, with the image of me pointing to the floor. We were in such a rush that I never knew what the meaning of that was."

"That's funny, Gram. When my students and I repelled down into the crystal city, it became obvious that we would have a hard time returning to the surface of the cavern the way we went down. So, as I was investigating the inside of the crystal pyramid, I noticed a narrow tube going upwards. I took the risk of venturing up to see if it led somewhere, and at the top, I saw that pictogram of the family pointing down. When I brushed the soil away, I saw the hatch door and flung it open. It led to the side room where I found the toy box. If I hadn't seen that, we might have had to attempt to climb up the sheer walls. It really saved our lives. Grandfather must have known I would go that way to find the exit. Wow."

"That explains that curiosity for me. Now, for your question of how we escaped and where we went. You know of the volcanic tubes that are like a highway under the earth? That crystal city also had a tube on the back side of the pyramid plaza. We entered it by way of the tube slide down to the base and went into the volcanic highway tube. The G-force was so intense that Athena screamed almost all the way there.

"I think that's one of the reasons she never liked carnival rides. Anyway, we ended up in Egypt, inside the Great Sphinx. We stayed there for what seemed like days. The inside was a chamber with golden couches and a full ship at the end of the long corridor. There were treasures everywhere. That's where Skylar found a golden ring that she said called to her. I thought she just wanted it, but Seth said it had belonged to Osiris and that she was family, therefore, she could claim it. The story of that ring is that Osiris used it to access the underworld. That had always represented the afterlife, but Seth informed me that it was a ring used to connect with the parallel universe."

Gram went silent suddenly. She drank deeply from her tea and stood to walk around the lower yard. Jeff wasn't sure what had happened, but he didn't want to upset her. It looked like a memory had caused her grief, and he understood how that felt.

Jeff went into the house and made a plate of fruit and cheese, and refilled his tea. His grandmother was still walking around the yard, pacing and thinking. He called down to her, "I've prepared a snack. Maybe we just need a break for a few moments." He could see her rubbing her forehead and slumping, something he had never seen her do. Now he was beginning to worry about what she had to tell him next. "I'll be there in a few moments. Why don't you just relax?" Gram said softly.

"I wasn't aware that anything was inside the Sphinx," Jeff asked as his grandmother finally came and sat on the patio chair. He moved over to the other cushioned chair and placed the plate next to them on the adjoining side table.

"You're forgetting that was circa 1906. It was pillaged not long after that, which is a shame as they couldn't figure out the reason for its existence for eons. The ship was able to set afloat in the moat that surrounded the sphinx and the Giza plaza and joined the Nile. Kind of a king's boat dock, if you will. At any rate, after our stay there, we were greeted by what appeared to be another Annunaki elder. He informed us that due to the speed in which we arrived the ship to Sirius was not ready. We would be able to wait in the King's chamber until it arrived. It would take about nine days."

"Why not use those pendants, wouldn't they have transported you there?"

"Sirius is not connected that way. It is the only portal to access the time rifts we had used when we left Alpha Centauri to this time and place. We must travel there to return to our current timeline. It was going to be a long journey."

"Did something happen when you attempted to move through time? Was that when things changed?"

"No, let me complete the story for you. We followed the Annunaki through underground tunnels and up to the King's chamber through a vertical lift that used forced air to transport us. When we arrived, it wasn't a Holiday Inn, or even a place to lay our heads. It appeared to be a power plant with what looked like lasers emanating out of the opening. I was very concerned for the children. We bypassed that area into what looked a lot like our domicile in the cavern at the Grand Canyon. It actually looked like they had just transported everything as it was straight into this room. The walls were decorated in colorful hieroglyphics showing Osiris, a boat that looked like a portal, and other magnificent artwork. We sat down and had dinner. We slept as if we were home."

Jeff stopped to eat some of the fruit, and Gram joined him. She got up and grabbed the pitcher of tea, and refilled their glasses. It was going to be a long explanation, Jeff thought to himself.

"Going on, Skylar had taken the time to travel to all the areas that the Annunaki allowed her to go. It was on the third day there that your grandfather came to me and sat me down to tell me that Skylar's destiny was imminent. He looked terrified. That made me very concerned. I thought maybe this was the end of the world as we knew it. But did that mean that all humanoids would be eliminated or illuminated or what? Seth simply said, 'Things will never be the same.'

"Skylar finally came to me and asked me to come with her. She had found a section that had a story about her 4th Great-grandfather Osiris and she said that the ring she had was actually a key. It is the key to the portal that brought Osiris in and out of the parallel universe. I followed her deeper into the different compartments inside the Great Pyramid, and we finally came to a rather small room with the hieroglyphic instructions on the use of the key. It came with a warning that stated that the other world was a negative energy world and that travel there could be catastrophic. I told Skylar to talk to her father about the room and what the key represented. I was concerned because, just like in Sedona when Skylar wanted to visit the vortex, she was told that some energies were bad for her.

"Seth sat down with us that evening and we talked about the parallel universe. The warning was real. Because Skylar was born in the positive energy universe, positive and negative energies can cause chaos. Seth asked Skylar to wait on any other adventures or misadventures until he could work out what the outcome might be. Osiris was a different sort of alien, having been born on a planet in the Orion nebula. It was possible he could move between positive and negative energies by changing his frequency."

Jeff asked if this was part of the omniverse. He knew of the other universal positions and that as far as he knew the negative universes were off-limits for the very reason his grandfather had mentioned. It was the afterlife, and the solid bodies lived on the positive universes and the ethereal lived on the negative universes. Maybe everything he knew was in question at the moment.

Gram looked very sad as she continued with the story. "Seth spoke to me that night. This was the destiny that was foretold by the mystics and the Federation. Our daughter, the universal child, was to use that ring to enter the underworld. There she would find her negative self. When they approach each other and touch, both worlds would be altered. He told me she would not return from this, Skylar of that world and Skylar of ours would cease to exist. I asked what he meant and he just lowered his head. It is the end."

"What do you mean it's the end?" I asked. "We are still here. My mother and uncle are still here. You still exist. That isn't the end. I never heard of a great aunt named Skylar, but then again, didn't you try to stop Skylar from using the ring? I know you talked about time-altering events, but wasn't that exactly what she was about to do?"

"I wanted to, but Seth said that destiny had to be played out," came the response. "The next day, I held Skylar in my arms until she pulled away, telling me that was enough love. She was on a mission of discovery. She told me she loved me and went off to her destiny. Seth and I sat in our domicile and wept. The twins were completely unaware of what was about to occur. About 20 minutes later, we heard an implosion sound. The air rushed past us and out towards the edge of the earth, then rushed back in. It was what a nuclear bomb would look like when exploding."

"Did it feel different to you?" I asked.

"Oddly, it seemed to be the next morning, and the twins didn't seem to miss their sister at all. It was as if they didn't remember they ever had one. Later, the transport came and we were off to Sirius and to our own timeline. I didn't really recognize the society as it was presented to me. Seth and I had full memories of the old timeline and the new. The world you were born into was well controlled by the Federation of Worlds and the Smithsonian who kept a tight lid on any mention of the old world. I always wondered how they explained the war wounds from the many veterans, but they probably just disappeared them like the history that happened before the creation of the new universe, the new earth.

Although this existence is peaceful, tolerant and a utopia compared to the world I grew up in, it is still hard for me to swallow the complete erasing of history by those record keepers. Until then, I hope the knowledge of the what was and the what is of today away from the public. This is why they are so cautious of your archeological expeditions; the old past is still available for discovery if you don't automatically bury it.

Welcome to Earth, the paradise not lost and the future guaranteed. It could have been just as perfect in the other reality if humans weren't so greedy and intolerant of others' belief systems or way of life. We could have made a difference, I think. Maybe, in the future, you will find that pearl of wisdom to bring history back to the conscious minds of everyday people. But, until then, I hope you can live with the knowledge of what was and what is today. Because it was the end of the world as I knew it."

# Star Map

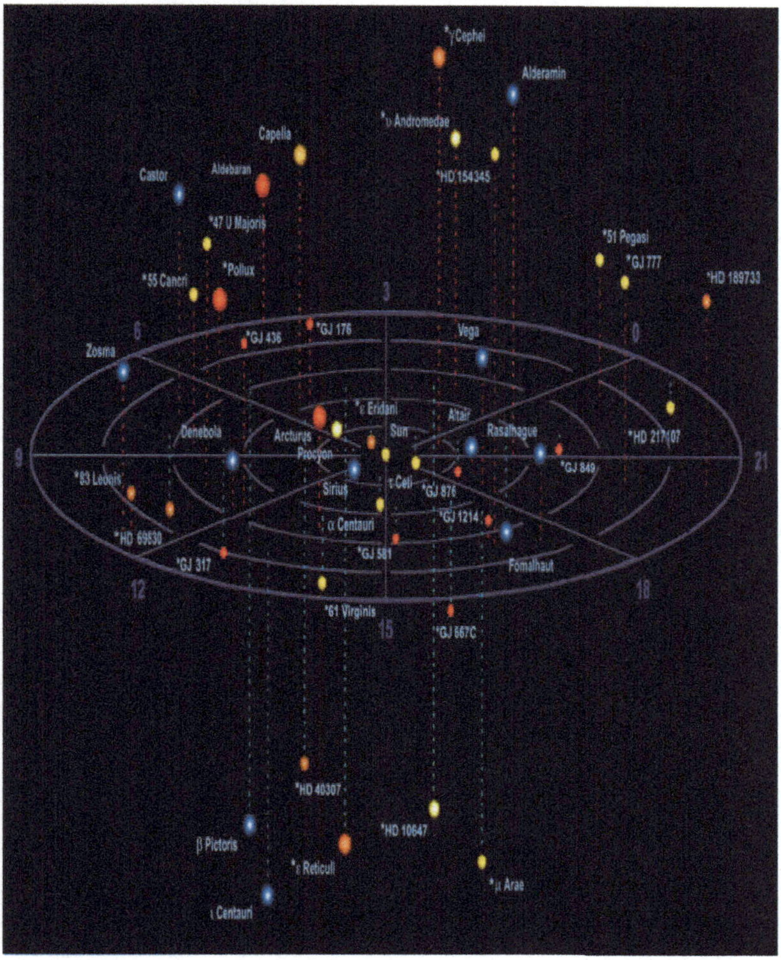
*γ Cephei
Alderamin
*υ Andromedae
Capella
*HD 154345
Castor
Aldebaran
*47 U Majoris
*51 Pegasi
*Pollux
*GJ 777
*55 Cancri
*HD 189733
3
*GJ 176
6
*GJ 436
Vega
0
Zosma
*ε Eridani
Altair
Denebola
Arcturus
Sun
Rasalhague
*HD 217107
Procyon
9
21
*GJ 849
*83 Leonis
Sirius
τ Ceti
*GJ 876
*GJ 1214
α Centauri
*HD 69830
*GJ 581
*GJ 317
Fomalhaut
12
18
*61 Virginis
15
*GJ 667C
*HD 40307
*HD 10647
β Pictoris
*ε Reticuli
*μ Arae
ι Centauri

# GLOSSARY

**Alpha Centurion Beings:**

Light body, humanoid, vibrational elements create an almost transparent image, they have a violet aura.

**Alpha Draconian:**

Beings who perceive Earth or Terra as their home world, not a friend to humanoids, but members of the Galactic Federation. Shape Shifters. Some Draconians are from the Draco Constellation and believe they terra formed the planet Terra (Earth) many millennium ago. They are part of the Evil Six who wish to genocide all humanoid species out of the Universe.

**Anakim:**

Els or El-der race referred to in biblical text as Elohim. This species are considered by the Universe as the original humanoid species. Recorded history has no specific location as to their home world or even their Universe but history reports that they appear in early archives as being Lyran.

**Annunaki:**

Humanoid race reported to be the creators of modern humans in the ancient Sumerian Cuneiforms. They are reported to be from our own Sol Star system, the 12th planet

Niberu {Zecharia Sitchin} also called Marduk which is on a long elliptical orbit around our sun (Sol). Niberu orbits close to Terra (Earth) once every 3,600 years. These humanoid are said to have come to Terra 450,000 years ago to mine gold to preserve their atmosphere. Around 300,000 years ago the commander, Enki exploring Terra convinced his brother Enlil a genetic scientist, to genetically alter the indigenous creatures to produce higher thinking hybrid to assist with mining operations.

Hopi tradition claim the Annunaki as their saviors during the great flood. The Hopi elders called Ant Friends (anu meaning ant and naki meaning friend) by the legends, but in truth the protective suits worn by the pilots had large protective eye covers and cap antenna which they used as communication with those who had gone underground prior to the second extinction event to hit Terra. Current translation is People of Anu

**Andromeda:**

Peace loving humanoids who have transcended from the third dimensional physical body to the fourth dimensional light body. They appear to be translucent and their skin appears slightly blue.

**Arcturian:**

Their planet Arcturia, 36 light years from Terra, is in the Bootes Galaxy. Arcturians are humanoid, tall between 7 to

9 feet tall, light almost white skin transparent skin, almond eyes, normally dark but variety exists. They are telepathic, with the power of telekinesis, advanced healers, highly enlightened teachers and mentors.

**Amphibians:**

These so called aliens are in fact humans (Pleiadians/Vegan/Lyran origin) who over many millions of years have become part of the sea, often referred to as sea nymphs or mermaids. These amphibians' are part of mythologies of Atlantis . When the "Wall of Fire" decimated the Sol star system these Lyrans went underground, under sea or out of the galaxy. Natural evolution have created these sea nymphs and mermaids.

**Atlans:**

Mythology has Atlans or Atlantians as Titans. These humanoid were considered gods and goddess' but ego and conflict between the inhabitants of Atlan and Mu created conflict and war, then nature overtook Terra and the climate change created the Great Flood. Mu's inhabitants currently referred to as Telosians. They went deep within the earth to seek shelter in the volcanic tubes later creating great cities (these aliens have been seen in the Mount Shasta, California area where one of many entrances to the underground highway systems within the volcanic tubes exists. **Telos** in

North America and sister country is **POSID** in South America are two of the Atlantians and Lemurian realms.

**Blue People:**

**Cherokee** referred to them as blue people or as moon people. They also share mythology in India and Tibet (see Nagas).

**Cetians: (Tau Ceti)**

This humanoid species is said to have tan skin, pointed ears and elongated heads. Tall, lean muscular bodies, angular faces with large eyes. Think Egyptian Gods. Originally from the Lyra. Early explorers, colonized habitable planets in the Cetus Constellation. They are also referred to as Gizan.

**Culhua:**

Hubrid race Reptilian/ Dragon or serpent race spliced with El-der. These hubrids originally appeared as a red dragon blend with human features, red scaly skin and serpent eyes. They had large skin flaps to expend for gliding and flight. No mythology could be found on this race but references to the red dragon appear in stories of China as well as South America.

**Dinosaurian: (Saurian)**

Hairless, green-skinned reptiles with bulging skulls, luminous eyes which are cat-like with three fingers. Modern description is humanoid features with slightly longer noses, skin is greenish to blue. Hair is course, body type is slim, tall with strong legs…athletic. Eyes are a combination of gold, blue or green with pupils that are slightly elongated much like a cat. Hair is a variety of shades from light to dark. They have the ability to be chameleon in nature and can shape shift making identification almost impossible.

*There are factions within this group that are members of the Federation who assist in the protection of humanoid life, others have joined the evil six.

**Dracos: Mothmen-**

Dragon race from the constellation Draco. Reptilian winged lizard race part of the Saurian species Members of the Evil Six This species claims to be the original inhabitants to Terra therefore feel that humanoids have taken their indigenous home.

**Elohim**:

El-der race is known as the oldest humanoid species in our Universe. Their original planet or constellation or even the universe in which they came is unknown. They first appear in the archives as being part of the pre-historic records of Lyra. They are still at war with the Orion and Reptilian races who manipulated the Zeta Reticulians. Their history pre-dates all known documents. They were part of the mythology of early Terra being referred to as giants. They were reported to have red or blonde hair, muscular body types as well as some reptilian features shared by Pleiadians, wings). Biblical reference to the Elohim can be found in the Bible, Revelations 12:7, "A war arose in heaven, Michael and his angles fighting against the dragon and his angels. Probably the Great War seen by Terran historians recorded in both cuneiform and other world archives.

**Hybrid:**

A blending of human and alien DNA which is with other humanoid creatures not of Terra.

**Hubrid-**

A sliced species of reptilian and human DNA. The cross breeding with reptilians created a human looking shape shifter. They walk among us unnoticed as do the hybrid star children, who are Terran and human alien crossed beings. Both hubrids and hybrids contain a soul matrix.

**Janosians:**

Humanoid species who is currently attempting to seek asylum on Terra. They are working with both the secret government on the surface as well as the Telosians who control middle earth. They are outside the observable range orbiting. They are from the planet Janos whose atmosphere was decimated by nuclear implosions. This was in part due to a civil war. Neither group has welcomed them to date. They are considered a dangerous species.

**Lyran:**

Ancestors of the El-der race. This species has both humanoid and reptilian/draconian indigenous refugees. Lyra was made uninhabitable early in the Great War but may, at this time may be habitable.

**Naga:**

These alien are part of the Indian Mythology. They are reptilian/dragon, blue skins from Draco and Reticula. These reptilian/dragon aliens were not part of the evil six. These were thought to be Gods by early India people. They were scaled skinned, blue or green with snake hair and many arms. Artist interpretation was many times inaccurate and expanded from reality. They were scaled, bluish skin, large eyes, and humanoid physical features, both beautiful and frightening. Greek mythology had Medusa who very well may have been Naga, India had Vishnu in the god and goddess traditions around the globe.

**Nordics**

**An** all human species. They are most reported to be Arian with blond hair or red and large blue eyes. See Lyra, Pleiades, Vega.

**Omniverse:**

The four known parallel universes within the cosmos. They are matter, anti-matter, physical (comprising third and fourth dimensions), and ethereal (comprising fifth through twelfth dimensions). The concept of the Omniverse or Multiverse is the theory of alternate realities using String Theory. A multitude of bubble universes could have emerged from the primordial cosmic sea of energy.

**Orions:**

A generalization of a species when in reality the Orions are diverse containing both humanoid, reptilian, draconian and Zeta.

**Orion Nebula:**

Said to be a doorway to infinity. The watchers of this portal guard against those who would bring harm to any species. They are benevolent, peaceful enlightened watching the gate which is used much like a super highway. This gate transcends both time and distance and can traverse the Omniverse of which there are a known four…matter/anti-matter, physical/ethereal.

**Reptilian:**

already described in the Dinosaurian, Saurian, and reptilian species.

**Teros:**

A country, Teros situated under Mount Shasta in California formally known as the lost continent of Lemuria or Mu as it is referred to in ancient times. The great flood sank the continent which was off the Pacific Coast line of Europa. The original tribe of Lemuria live underground. Telosians are the inhabitants of the country Teros.

**Saurian:**

Oldest occupant of the planet Terra (Earth) in the Sol star constellation, having been on this planet for 64 million years. Originally part of the Stenonychosaurus race which have evolved from dinosaur to humanoid appearing species. Sauroids were part of the underground labyrinth of the Giza plateau. Highly advanced species which built great cities around the world as well as using the inner sea (currently known as Death Valley) for trade with the Lumarian races. These species is part of the evil six Draco Consortium.

**Star Seeds:**

Earth born humans who have a shared soul matrix with alien life.

**String Theory:**

Einstein was partially correct. Everything in existence is made up of particles down to the smallest quarks, under those are dancing filaments of energy that resemble vibrating **string**. These filaments all are made up of one entity. Call this the divine or the ONE. It can be viewed as the original substance. Everything is made of the One.

.

# TIME DANCING, A NOVEL WRITTEN BY PAULA A ADKINS

## Synopsis

Professor Jeff Ryan, of the Smithsonian Archeology Research and Exploration Unit, gathered his brightest students from the University of Arizona for an exciting investigation of the cave dwelling within the South Rim of the Grand Canyon. Due to erosion, what appeared to be an effigy of the god Ra had been seen near the stone cut opening to an ancient cliff dwelling. This area had been off limits to everyone worldwide for more than one hundred and forty years.

The Smithsonian agreed to allow this expedition, providing anything which might prove to be a time disturbance would be left where it lay, and that the information would be forever held in confidentiality. It was a strange request, but the Professor agreed.

Within the rim of the Grand Canyon, beyond the cave dwelling, a great crystal citadel was found. Evidence of a vast spiritual library, which included Aztec, Mayan, Egyptian, and Indian deities, as well as hints of starships and a galactic war, was found among ancient relics.

Oddly, hidden within the dig, was a mysterious twentieth-century journal which mentioned a violent past not known in the universe he lived in. He felt connected to the writer of this journal. He did not know why, but he had to hide this evidence of a potential time disturbance until the mystery could unfold. The author of this diary or journal felt somehow familiar, like family, but she wrote of a place that was not the Earth he knew. It was a profound mystery, a hint of a different timeline. It became his passion to discover this time anomaly for himself.

# ABOUT THE AUTHOR

# PAULA A ADKINS

Paula grew up in El Paso, Texas, and spent many days with family and friends investigating UFOs and the myths that were available about the activities in the nearby White Sands Missile Range. Science became a passion early on in her life. She attended the community college in San Diego, California, and then went on to the University of Phoenix to obtain her Bachelor of Science degree. She worked for both the San Diego Unified School District and the Scottsdale Unified School District, exhibiting a talent for special education.

She married Stan Adkins, a noted musician, and enjoyed a loving life with him for thirty-eight years until he succumbed to cancer. She began writing her novel years ago, but as often happens, life sometimes interfered with the plans she had made.

Paula's novel was inspired not only by her own life events, but also by an unpublished novel written by her parents back in 1962. This unpublished work was a science fiction tale about the so-called gods of Atlantis who, according to their novel, were aliens from a dead planet, Titan, which is where the asteroid belt lies today. Their main characters were Diana and Apollo, who you will find in Paula's novel.

The upcoming novel will be a sequel entitled "War of the Titans." It was written by Pauline and Hubert Gay, now deceased, and is currently being rewritten by Paula Adkins.

Made in United States
North Haven, CT
10 November 2026

22084533R00183